Leandro Barros Oliveira

Agriculture and Forests in Sumidouro (RJ)

Leandro Barros Oliveira

Agriculture and Forests in Sumidouro (RJ)

Implications of the New Forest Code for the Campinas Watershed

ScienciaScripts

Imprint

Any brand names and product names mentioned in this book are subject to trademark, brand or patent protection and are trademarks or registered trademarks of their respective holders. The use of brand names, product names, common names, trade names, product descriptions etc. even without a particular marking in this work is in no way to be construed to mean that such names may be regarded as unrestricted in respect of trademark and brand protection legislation and could thus be used by anyone.

Cover image: www.ingimage.com

This book is a translation from the original published under ISBN 978-613-9-65482-6.

Publisher:
Sciencia Scripts
is a trademark of
Dodo Books Indian Ocean Ltd. and OmniScriptum S.R.L publishing group

120 High Road, East Finchley, London, N2 9ED, United Kingdom
Str. Armeneasca 28/1, office 1, Chisinau MD-2012, Republic of Moldova, Europe
Printed at: see last page
ISBN: 978-620-7-77980-2

Summary

Thanks

First of all, I would like to thank everyone who contributed directly to the success of this research: my advisor Paulo Alentejano; the members of the qualification and defense committees: Douglas Pimentel, Luiz Moraes and Eduardo Stotz; all my colleagues in the master's program; the entire team of the Post-Graduate Program in Science, Environment and Society Teaching at UERJ; my colleagues at the local Emater-Rio office: Jader Campanati, Priscila Barbosa and Ricardo Belo; to my regional supervisors: Alexandre Jacinto and Marcos Belo; to the former Secretary of Agriculture and Environment of Sumidouro: Landirlei Gomes; to Marcelo from the Centro de Documentaçâo Histórica Pró-memória Sumidouro; to Jaqueline and Moabe from the municipal library; to Geisa Diniz, Sandrianne Gripp, Professor Sueli, Marcelo Ezequiel and Anderson Oliveira who helped me at different stages of this project.

I would also like to thank all those who helped pave the way for me to get here: my teacher Célia Maria, who changed the course of my life from grade 8^a ; my friends and teachers at Colégio Agricola Nilo Peçanha, for the maturity and experience provided; my colleagues and teachers at CEDERJ, especially my teacher and friend Anderson Portugal, who encouraged me and gave me all the support I needed to apply for the master's program; my wife Kênia Lima for her support in difficult times and my parents Tânia and Vladimir for encouraging me to study.

Preface

The book that Leandro Barros Oliveira is offering us is not just the result of his Master's thesis, which I had the honour and pleasure of supervising, but the materialization of years of work with family farmers in Sumidouro. It was with the attentive and concerned eye of an Emater-Rio technician in Sumidouro that Leandro looked into the difficulties faced by family farmers in this small and undervalued municipality in the state of Rio de Janeiro.

Sumidouro is one of the main horticultural centers in Rio de Janeiro and is responsible for an important part of the supply of products such as tomatoes, lettuce, cauliflower, peppers and other products to the metropolitan region and the capital. Most of this production is carried out in smallholdings using family labor, although there are also daily wage earners and partnership and leasing practices. The use of pesticides in this production is also very significant, practically a requirement of the market which demands large, showy fruits and leaves from producers that can hardly be produced without the use of chemical products. Given the small size of the land used by most farmers, intensive use is also common, which generally leads to the use of areas with steep slopes and riverbanks, practices that are contrary to environmental legislation.

It is this problem in particular that Leandro focuses on, analyzing the changes introduced in environmental legislation through the New Forest Code and the adaptation to it by family farmers in Sumidouro. Using a variety of research techniques such as questionnaires and focus groups, Leandro analyzes the contradictions between agriculture and environmental conservation in the municipality.

Leandro takes a critical look at the process of changing the Code, pointing out that its drafting fundamentally served the interests of agribusiness (the association between large landowners and big agro-industrial capital with the financial and legal support of the state, in Guilherme Delgado's felicitous definition), Leandro points out how the small producers of Sumidouro reproduce the dominant ideology regarding the relationship between agriculture and environmental conservation, but they also point out the central responsibility of the latifundia in the process of environmental devastation in Brazil.

Finally, Leandro points out the limitations that exist today for overcoming these contradictions between agriculture and environmental conservation, given the lack of support for the development of agroecology as an alternative to the conventional model. In his words, "the implementation of public policies that make the agroecological transition viable on family farms is the main alternative to the problem at hand, and these policies need to cover the entire production chain, guaranteeing credit, encouraging production and facilitating marketing. In addition, education and systematic monitoring of family farmers must be guaranteed, since the great challenge is to effectively raise awareness and this ideal only has full meaning outside the conventional system, in a logic where industrial capital does not dictate the rules of production systems."

This is the challenge to which Leandro invites us to reflect on the research he carried out with the farmers of Sumidouro and which is materialized in this book.

Happy reading, everyone.

Paulo Roberto Raposo Alentejano

PhD in Development, Agriculture and Society from UFRRJ Professor at the Faculty of Teacher Training at UERJ

Presentation

The municipality of Sumidouro, located in the mountainous region of the state of Rio de Janeiro, has atypical characteristics when compared to other municipalities in the state. More than half of its population lives in rural areas and its main source of income is family farming. Among the various municipalities, the Campinas Hydrographic Microbasin (MBH Campinas) stands out as the most representative in demographic and productive terms.

Family farming in MBH Campinas is characterized by the intensive use of soil, pesticides and other external inputs. Over the decades, this farming model has contributed to the reduction of a large part of the areas of native vegetation in the watershed and even today puts the remaining forests at risk.

I was born in Barra Mansa and moved to Sumidouro in 2010, when I was approved for the position of Rural Development Agent at Emater-Rio. In the first few years of my professional experience, I was able to observe the various consequences of this production model: I saw gullies[1] appear, springs dry up, rivers being diverted, many forest fragments grated until they disappeared and I heard about cases of pesticide poisoning, some of which unfortunately resulted in death.

During this time, my colleagues and I achieved very few results in our countless attempts to encourage the adoption of conservation practices. That's when I began to realize that there was a lack of systematic studies to guide the planning of the rural extension service[2] in the region. All this gave me a strong sense of unease that catalyzed my turn towards research.

Thus, motivated by the controversy surrounding the reform of the Forest Code, between 2014 and 2017 I carried out the research that gave rise to this book. The aim of this study was to investigate the relationship between family

---

1Geological phenomenon consisting of the formation of large erosion holes caused by rainwater and bad weather where the vegetation no longer protects the soil. The gully makes the soil poor, dry and infertile (FERREIRA, 1986).
2Rural extension: more information from the second paragraph on page 25.

farmers in MBH Campinas and the forest remnants on their properties and, based on this, to carry out a critical analysis of the implications of this new law, bearing in mind the socio-environmental adversities caused by the model of agriculture practiced in my study area.

I think it's important to point out that I don't intend to exhaust the subject, but I hope that this research will help us to reflect on the praxis in favor of a more solidary agriculture with less harmful impacts on humanity and the environment. Furthermore, I hope that the knowledge generated by this publication will be useful to the municipality, the rural extension service and any other organization that works with producers in the watershed.

Happy reading!

Leandro Barros Oliveira

The author

Introduction

The proposal to reform the Forest Code had been on the agenda of the National Congress since the 1990s, and was the subject of intense debate between opposing social forces. On one side were environmentalists, social movements and peasants, and on the other the ruralists, supported by landowners and rural businessmen. Over the years, various members of parliament linked to agribusiness have joined together to form what we know today as the ruralist caucus, resulting in greater political bargaining power for the development sector. As a result of this arrangement, on May 25, 2012, Law No. 12.651 was published, a reformulation of the 1965 Forest Code. The new law was considered controversial because it disregarded the technical-scientific concepts applicable to preservation areas and granted amnesty for environmentally damaging conduct perpetrated before July 2008. At the time, the issue had great repercussions, generating much controversy and intense debate in universities.

Even though the new law brought significant changes for small properties by drastically reducing the obligation to restore protected areas[3] , Sumidouro's rural producers remained indifferent. In fact, understanding why this was the case was easy: the previous Forest Code, although very restrictive and theoretically making much of the agriculture practiced in the municipality unviable, had never been effectively applied in the region. The government had no effective mechanisms for enforcement and landowners in general, even without complying with the law, were able to finance, divide and sell their land without any major problems. So for many, nothing had changed.

In 2014, with the implementation of the Rural Environmental Registry (CAR[4]) and the publication of the decree regulating the Environmental Regularization

---

3Protected areas: concept discussed on page 33.

4The CAR is a tool provided for in the new Forest Code, which consists of a mandatory georeferenced declaratory register for all rural properties and allows properties to be monitored by satellite, thus making it easier to control deforestation.

Program (PRA[5]), the subject once again resonated among agricultural professionals. From that moment on, all rural property owners and squatters who wanted to share or sell their land and access rural credit lines had to register with the Rural Environmental Registry (CAR) via the electronic address.

At this time, the reform of the Forest Code and the obligation for producers to register with the CAR were widely publicized in MBH Campinas through meetings of the Rio Rural program[6] held by Emater-Rio and training sessions promoted by the Sumidouro Rural Union in partnership with the National Rural Apprenticeship Service (SENAR). As a result, many landowners, fearful that their properties would be restricted and that they would not be able to renew their agricultural investment and funding operations, became aware of the issue.

In this context, I was faced with an ambiguous situation. At the same time as the loosening of the Forest Code was seen as controversial and questionable from an ecological point of view, rural landowners were finally becoming aware of their responsibilities, albeit minor ones. In MBH Campinas, the implementation of the CAR and its link to rural credit, together with the mobilization promoted by the Rio Rural program, have, in a way, renewed expectations of a change in conduct in terms of conserving protected areas on rural properties. Something that never happened under the previous Forest Code.

These questions motivated my interest in investigating the relationship between family farmers and the remaining forests on their properties, with the aim of also looking at the possible impacts of the application of the New

---

5The PRA is a set of measures to guide the actions to be taken by rural landowners and squatters to regularize the environmental status of their properties. More information on CAR and PRA on page 49.
6The Rio Rural program, developed by the State Secretariat for Agriculture, Livestock, Fisheries and Supply, promotes sustainable development projects on rural properties. For more information, see page 54 or visit www.microbacias.rj.gov.br.

Forest Code in the watershed. In this way, I developed the research presented in this book during the period in which I was linked to the Postgraduate Program in Science, Environment and Society Teaching at UERJ.

This book is organized as follows: the first two chapters deal with historical aspects, political and legislative events and the main concepts related to the subject. Then, in chapter 3, I present the study area, based on data from Emater-Rio, the IBGE, the Municipal Department of Agriculture and the Environment and the empirical experience accumulated over eight years of working as a rural extensionist in the municipality in question.

In Chapter 4, I discuss the methodology used, which consisted of applying structured questionnaires to quantitatively survey the protected areas on the properties analyzed and conducting focus group interviews to obtain subjective information related to producers' perceptions. In Chapter 5, I present the results obtained, comparing them with the framework used in order to understand the questions proposed in this investigation and, finally, in the Final Considerations, I close the text by presenting the conclusions reached.

Capitulo 1

Historical aspects

1.1 Historical dimensions of the exploitation of the Atlantic Forest

In the 16th century, after the arrival of the Portuguese colonizers in the territory that would later be called Brazil, monoculture plantations began and, together with mining, timber exploitation and cattle ranching, it is estimated that between the 16th and 18th centuries, more than 30,000 km^2 of the Atlantic Rainforest were devastated (DEAN, 1996).

From the end of the 18th century, in addition to the areas already devastated by sugar cane, mining and beef cattle, the mountainous region of the state of Rio de Janeiro, which still had dense areas of forest, began to be targeted for the cultivation of coffee, which became the product of the large farms donated in sesmarias[7] and was the basis of the export economy of the state of Rio de Janeiro for decades (DEAN, 1996).

Since coffee is demanding of drained soils, it began to be cultivated on steep slopes and hilltops in the mountains of Rio de Janeiro, areas whose ecological function of the original forests was to stabilize and recharge the water table. At the time, it was believed that coffee had to be planted in "virgin" soil, a fact that intensified the practice of burning in order to make the nutrients from burning biomass available. Because coffee is a perennial plant capable of remaining productive for around thirty years, at the end of its productive cycle it was common for old plantations to be abandoned, as it

---

7Sesmarias were donations of land made by the Portuguese Crown to its agents and settlers in the process of "occupying" Portuguese America. The Sesmarias Institute was the colonization policy put into practice in Portuguese America during the reign of King João III, when the hereditary captaincies were created (SILVA, 2018).

was easier to clear new areas of primary forest to maintain production. As a result, the crops advanced through the highlands from generation to generation, leaving in their wake a landscape marked by bare mountains (DEAN, 1996).

The coffee trade fostered demographic growth on a scale never before experienced in the country. The human population of the southeastern region of the Atlantic Forest increased from one million people in 1808 to 6.4 million in 1890 (DEAN, 1996). However, the marketing of Brazilian coffee was essentially dependent on the nations that imported the bean from Brazil and had already been showing signs that it would collapse since the second half of the 19th century (DEAN, 1996).

It should be noted that there are important regional differences in relation to this process: coffee growing in Rio had been in decline since the second half of the 19th century due to the abolition of slavery and soil erosion (with the exception of some municipalities such as Sumidouro, as will be discussed below), while in São Paulo it expanded based on the labor of immigrant settlers, only declining with the crisis of 1929 with the "crash" of the New York Stock Exchange.

With the decline of coffee in the 20th century, Rio de Janeiro's economy experienced a new development model based on urbanization and industrialization, which further intensified the impacts on the Atlantic Forest. However, not all municipalities managed to adapt. The newly-emancipated municipality of Sumidouro, which had its economic heyday based on the coffee economy, faced a long period of decay, until this situation was reversed with the introduction of a new model of agricultural production, this time based on olive growing[8] (CENTRO DE DOCUMENTAÇÃO HISTÓRICA PRÓ-MEMÓRIA, 2016).

---

8 Olericulture is the cultivation of vegetables.

1.2 Breaks with the traditional: the arrival of olericulture and the modernization of family farming in Sumidouro

In the first decades following its emancipation in 1890, Sumidouro was still thriving economically, even when many of the surrounding towns and villages were already suffering from the abolition of slavery and the severe coffee crises. In 1885, the construction of the Leopoldina Railroad, which passed through Sumidouro, contributed greatly to maintaining economic growth during this period. This was possible because access to transportation facilitated the flow of coffee production. The municipal economy remained relatively stable until the mid-1920s when external pressures increased, making it impossible to perpetuate the coffee cycle.

Figure 1. downtown railway station in Sumidouro - RJ. Source: Pró-memória Sumidouro archive.

At that time, unlike in most coffee-producing municipalities, most of Sumidouro's rural population remained in the countryside. Some mills run by Italian and Swiss immigrants continued to produce sugar and rapadura, while corn, beans, potatoes and other subsistence products were grown by peasants.

According to records from the Centro de Documentaçâo Histórica Pró-Memória[9] , in the 1930s Sumidouro was already showing scars in the

---

9 Division of the Municipal Department of Education and Culture responsible for organizing the

landscape caused by abandoned coffee plantations. The area around the headquarters had been constantly burnt down, leaving several hectares of land unsuitable for cultivation. These areas were later used for extensive cattle breeding. However, the territory of Sumidouro still had regions that had been little explored, the so-called cold lands: areas of higher altitude, bordering the municipalities of Nova Friburgo, Teresópolis and Duas Barras.

Figure 2 - Sumidouro's headquarters in the mid-1950s. Source: Prò-memoria Sumidouro archive.

Due to the ideal climate and soil conditions, some farmers migrated to the "cold lands" and began growing oleric crops in these areas. According to Stotz (2012), the experience of Japanese immigrants who settled in the district of Dona Mariana in the 1960s was very important for the spread of olive growing in the "cold lands". According to reports gathered by the author in question, the Japanese rented farms and the surrounding peasants learned to work by watching them.

During this same period, in Brazil and in several other countries, a process of agricultural modernization was taking place, promoting radical changes in the way food is produced and marketed. These innovations were the result of the technological development that took place in the period after the Second World War, when rich countries began to encourage the adoption of modern

historical collection of the municipality of Sumidouro.

agricultural techniques in third world countries, under the claim that they wanted to put an end to hunger, since the cause of hunger was attributed to a production problem (GROSSI & SILVA, 2002). This paradigm shift became known as the "Green Revolution" and its product is what we now call conventional agriculture (ALENCAR *et al.*, 2013).

In conventional agriculture, diversified systems have been replaced by monocultures based on industrial inputs, such as: synthetic fertilizers, pesticides, artificial irrigation, soil mechanization, genetically uniform seeds

developed in laboratories, all with the aim of maximum productivity. According to Pereira (2012, p. 686):

[...] the millennia-old practical knowledge of the farmer himself has been replaced by scientific knowledge; local ecological cycles, based on endogenous resources, have been replaced by exogenous industrial inputs; the work that used to be carried out in coexistence with nature has been fragmented into parts - agriculture, livestock, nature, society - and each sphere has come to be considered separately, breaking the unity that exists between human beings and nature.

According to Heredina *et al.* (2010), the adoption of conventional agriculture or the "modernization of agriculture" was a process widely encouraged by the state during the military dictatorship. As Musumeci (1987) points out, the establishment of rural credit in 1966, the creation of the Brazilian Agricultural Research Corporation (Embrapa) in 1972, the Brazilian Technical Assistance and Rural Extension Corporation (Embrater) in 1974 and the launch of the National Plan for Agricultural Defensives (PNDA) in 1975, contributed greatly to strengthening this new paradigm.

The public service of Technical Assistance and Rural Extension (ATER) played a major role in disseminating the technologies of the Green Revolution. The ATER companies presented technological innovations to producers, including providing training and agricultural credit, with medium and large producers as the priority audience. This phase of ATER was called Productivist Diffusionism (BORDENAVE, 1985).

According to Egger (2010), the modernization of agriculture in Sumidouro began to be incorporated into production later than in other regions. The substitution of traditional elements, such as plows for tractors and organic fertilizers for chemical fertilizers, was more significant from the 1980s onwards, although some reports collected by Stotz (2012) show that in the 1970s, cargo buyers from Teresópolis were already carrying out married operations, inducing farmers to use pesticides:

[...] "There was a truck driver who used to take our entire load, from Sâo Lourenço, from Sumidouro, to Teresópolis to sell the load, just one truck! He brought the poison himself! The people who were going to take the loads also bought it" (Report collected by STOTZ, 2012, p. 120).

The use of technological innovations has led to a considerable increase in cultivated areas, while at the same time making farmers dependent on multinational companies that produce technological packages. In Sumidouro, adherence to conventional olive cultivation contributed to a return to economic stability. However, this commitment was so strong that today the municipality is among the biggest consumers of pesticides in the state of Rio de Janeiro (STOTZ, 2012; FUNDAÇÂO OSWALDO CRUZ, 2016).

Currently, the municipality is responsible for supplying vegetables to the metropolitan region of Rio de Janeiro via CEASA Irajà and still has part of its production sold through the Agua Quente Producer's Market in the municipality of Teresópolis. We can see, then, that even though the family-based workforce is still present, there has been a profound change in the organizational nature of the Sumidourense producer. Today, they are fully integrated into the market and operate in an essentially commercial way (EGGER, 2010).

1.3 The new ATER and agroecology

At this point, I think it's important to point out that after the end of this cycle of modernization in the countryside, the ATER service was no longer seen as something fundamental to the new dynamics of capitalism - after all, large estates had already been modernized and their owners had the resources to obtain private assistance. Thus began a process of weakening ATER, which began to be maintained almost exclusively by state governments after the extinction of the Brazilian Technical Assistance and Rural Extension Company (EMBRATER).

At the beginning of the 1990s, faced with this situation of low public investment, the role of ATER began to be rediscussed. In this way, ATER began to assimilate the foundations of the liberation pedagogy developed by Paulo Freire, incorporating the ideas of participatory planning and the construction of a "critical conscience" on the part of extension workers and farmers. Human and social development came to be seen as fundamental values in this phase of ATER, which became known as Critical Humanism (BORDENAVE, 1985). However, it was only in 2004, with the creation of the National Policy for Technical Assistance and Rural Extension (PNATER), that these ideals were incorporated into a public policy.

The PNATER established guidelines for ATER actions to implement and consolidate sustainable rural development strategies and stimulate income generation, with family farming as a priority (MDA; ASBRAER, 2012). According to Article 2 of Law No. 12.188 of January 11, 2010, ATER is now defined as:

A non-formal, ongoing education service in rural areas that promotes management, production, processing and marketing of agricultural and non-agricultural activities and services, including agro-extractivist, forestry and craft activities.

This new way of doing ATER has been catalyzed by numerous public policies that have emerged or been adapted since then, but obviously the hegemonic model of production and commercialization has brought and continues to bring numerous bottlenecks to the effective implementation of PNATER.

Contrary to this system, there are producers who work with the so-called agroecology, which is a science that seeks to combine the recovery of traditional knowledge with the application of the fundamentals of ecology in agriculture (GLIESSMAN, 2000).

Agroecology cuts across different philosophical lines[10] . Each one has its own particularities, but they all have in common the search for the efficient use of energy in the system, reduced dependence on external resources, the conservation of renewable resources and the production of healthy food (CERVEIRA and CASTRO, 1999; GLIESSMAN, 2000). According to Carrol *et al.* (1990), agroecology is an emerging science, formed from four areas of knowledge: agriculture, ecology, anthropology and rural sociology.

In agroecological cultivation, pesticides and synthetic fertilizers are abolished and maintaining high productivity depends on the dynamics of the soil's organic matter (MIYASAKA *et al.,* 1997). This requires the adoption of agricultural practices based on knowledge of the processes that occur naturally in the environment (GLIESSMAN, 2000).

According to Alencar *et al.* (2013), agroecology goes beyond the simple objective of producing without using pesticides, and also influences labor relations, the production chain and the health of rural people. To this end, agroecology advocates the use of low-impact practices such as no-till farming, crop rotation and intercropping, rationalized irrigation (drip or micro-sprinklers), the use of organic fertilizers, the use of mulch, the use of green manure, the rational management of spontaneous plants, the use of natural enemies, the use of biofertilizers and auxiliary mineral fertilizers with low solubility, as well as many other practices (SOUZA, 2000).

Municipalities such as Petrópolis, Teresópolis and Nova Friburgo are benchmarks in agroecological/organic production in the state of Rio de Janeiro. Although agroecology represents a small percentage of the total

---

10The main philosophical lines of agroecology: organic farming, biodynamic farming, biological farming, ecological farming, natural farming and permaculture (EMBRAPA, 2005).

production of these municipalities, the mountainous region of the state of Rio de Janeiro is one of the pioneers in this type of farming (FONSECA, 2009). Despite this emerging regional vocation, Sumidouro, which is also among the main food-producing municipalities in the mountainous region, is a long way behind the other municipalities mentioned in terms of agroecological production. According to information from the local Emater- Rio, until 2017 there were only three producers in the municipality with certification for organic farming, none of whom live in MBH Campinas.

1.4 From peasantry to family farming

The word peasant[11] , in its essence, points to a deep relationship of attachment to the land and autonomy in decisions about production, marketing, tradition, values and culture (COSTA and CARVALHO, 2012). With the arrival of modernized agriculture, peasant production, which had remained resistant to the assimilation of industrial inputs, was reduced to small self-assertive groups, almost always organized in social movements, in the fight for rights, given their competitive imbalance in relation to the capitalist agriculture that was being imposed at that time (COSTA and CARVALHO, 2012).

In this post-green revolution context, the term "peasantry" began to lose its relevance in public policies and was replaced by the generic nomenclature "family farming". In a general definition, family farming can be understood as the production of crops and/or livestock in which the family nucleus is responsible for the labor and management of the enterprise (PORTUGAL, 2004). For Abramovay (1992), the concept of family farming points to the socio-economic transformations brought about by the Green Revolution and the consequent deconstruction of the peasant profile, as we can see in the excerpt below:

11Peasantry: group of peasant families existing in a territory (COSTA and CARVALHO, 2016 p.23).

However, Lamarche (1993) understands family farming as a generic concept that incorporates multiple specific situations. The author understands that not only the modern farmer inserted into the market is embraced by the concept, but also the traditional peasant. Neves (2007) agrees with Lamarche (1993) and states that the term family farming refers to an ideal pattern of integration of a heterogeneous mass of rural producers and workers. In this synthesis - peasants, small landowners, tenants, partners, settlers, sharecroppers, rural settlers and landless workers have gained legitimacy as a category, becoming the object of specific public policies, different from those aimed at agribusiness companies.

It is important to note that the concept considered as the basis for this research assimilates family farming as a category supported by the National Program for Strengthening Family Farming (PRONAF), instituted in 1996 by Decree No. 1946 and updated by Law No. 11.326 of July 24, 2006 (NEVES, 2007; STOTZ, 2012).

In the Agricultural Census carried out in 2006 by the IBGE, 4,366,267 family farming establishments were identified in Brazil. The sector is currently responsible for 60% of the food produced for the basic diet of the entire population (SOUSA *et al.,* 2006). This data highlights the importance of family farming as an object of study, given the sector's great influence on human nutrition and the strong environmental impact caused by the means of production in vogue.

Capitulo 2

Legal aspects

2.1 **Reflections on preservation and conservation**

Porto-Gonçalves (1996) was categorical in proposing that the concept of nature is not natural, but is in fact a socially constructed concept that changes according to the historical and cultural context of each place and time. In this way, nature has been understood with man included and excluded, and today both visions coexist. A good example is the common expression: "mastering nature", which only makes sense if we consider that man is excluded from the concept.

The understanding of man as an entity dissociated from nature gave rise to the view that he is an enemy of the environment, thus the myth of untouched nature arose, an idealization of the wild world protected from human contact. In this way, the concept of preservationism emerged in the 19th century, along with the idea of isolating natural areas from human action. The creation of Yellowstone National Park in the United States was the milestone in a movement that spread to third world countries, giving rise to the parks that are now categorized as fully protected conservation units under Brazilian law (DIEGUES, 2001; MEDEIROS, 2003).

For the most radical preservationists, fully protected conservation units are the only areas that are really protected. However, the exclusion of human populations from natural spaces has proved controversial throughout history, since many traditional communities that have perpetuated themselves for centuries without causing major impacts would be left out of this paradigm (DIEGUES, 2001).

As a result of this ideological dichotomy, the conservationist current has emerged as a middle way that seeks to maintain nature with man included, acting in the environment in a rational way to ensure that future generations can also enjoy natural resources. This is based on principles such as: reducing the use of raw materials, the use of renewable energies, changes in consumption patterns, social equity, respect for biodiversity and the inclusion of environmental policies in the economic decision-making process (DIEGUES, 2001).

According to Medeiros (2005), the terminology "Conservation Units" has currently been misinterpreted as a matrix concept. However, the term "Protected Areas" covers a broader spectrum of typologies including: Conservation Units, Permanent Preservation Areas, Legal Reserves, Indigenous Land and Internationally Recognized Areas. This conceptual confusion is mainly due to the fact that there is no law that systematizes all the typologies and categories of protected areas in a single text.

Conservation Units are currently subject to the National System of Conservation Units (SNUC) through Law 9.985/2000 and APPs and Legal Reserves are subject to the Brazilian Forest Code (Law 12.651/2012). According to the author, "Protected Areas are territorially demarcated spaces whose main function is the conservation and/or preservation of natural and/or cultural resources associated with them". The concept encompasses both categories of integral protection and those that allow the sustainable use of natural resources.

Medeiros (2005) points out that the 1965 Forest Code was the most important instrument created for the protection of areas of ecological interest, as it objectively defined the bases for territorial protection of the country's main forest ecosystems and other forms of natural vegetation. However, there was no concrete strategy for enforcing the law and curbing deforestation on private property. As a result, the implementation of Conservation Units became the "apple of the eye" for many environmentalists and for a long time the Forest Code was left in the background.

The previous versions of the Brazilian Forest Code (Decree 23.793/1934 and Law 4.771/1965) clearly pointed towards a preservationist concept. The laws (with the exception of amendments that were included later) did not support the idea of sustainable use of natural resources in Permanent Preservation Areas[12] (APP) and Legal Reserves[13] (RL) and there was little differentiation in the rules for small and large landowners, which made it very difficult for small landowners to adapt. As such, many considered that reformulating the Forest Code was a necessary measure to guarantee the viability of family farming (OKUYAMA *et al.*, 2012). However, the rediscussion of the agenda was also convenient for the interests of landowners and large rural businessmen. This issue will be discussed in more detail in the following subchapters.

2.2 The Forest Code historically in conflict with conventional agriculture

The first government initiative to protect forests on private property took place in 1934 under the Getúlio Vargas government. In the preceding decades, coffee production had been a major factor in the national economy and cultivation areas were being expanded in a way in which the state had no control over which lands were public and which were private. At this time, the country's energy matrix was still based on the exploitation of coal and firewood, and the replacement of forests with agricultural areas put the stock of raw materials at risk. For this reason, in order to avoid an energy crisis, the first Brazilian Forest Code was created in 1934 (SELBACH, 2013; GARCIA, 2012).

12Permanent Preservation Area - APP: protected area, covered or not by native vegetation, with the environmental function of preserving water resources, the landscape, geological stability and biodiversity, facilitating the gene flow of fauna and flora, protecting the soil and ensuring the well-being of human populations (LAW No. 12.651/12, Art. 3, item II).
13Legal Reserve: an area located within a rural property or possession, delimited under the terms of Article 12, with the function of ensuring the sustainable economic use of the rural property's natural resources, assisting in the conservation and rehabilitation of ecological processes and promoting the conservation of biodiversity (LAW No. 12.**651/12**, Article 3, item III).

The aforementioned law, created by decree N° 23.793, made it compulsory for landowners to maintain a quarter of the area of their property in woodland (the so-called quarter) and also introduced the concept of "protective forests" which, according to Article 4 of the law, were those portions of the property that had, among other duties, the function of conserving the water regime, protecting rare species of fauna and guaranteeing the stability of areas at risk (steep slopes, dunes). This initiative was the first Brazilian legal mechanism related to environmental preservation. However, the 1934 law did not contain any further specifications or details, which made it difficult to apply.

In 1965, through Federal Law 4.771/65, the Forest Code was reformulated, giving a new meaning to the legal figure created by the 1934 Code. The term "protective forests" is no longer used, and the areas to be protected are now treated as Permanent Preservation Areas (PPAs) and Legal Reserves (LRs). The APPs were specifically classified as follows: marginal strips of rivers and lakes, radii of springs, the upper third of hills, restingas, mangroves, the edges of tablelands, plateaus, mountainous areas with an altitude of more than 1,800m and on slopes with a gradient of more than 45°.

The RL replaced the old "fourth part" to be preserved on properties, with the proportions for preservation redefined according to the Biomes. In the initial text, in the Amazon, 50% of the area of the property had to be preserved with native forest (later changed to 80% through Provisional Measure No. 1,511 of 1996), while in the rest of the country only 20% was required (in 2001 the area required for preservation in the Cerrado was increased to 35% through Provisional Measure No. 2,166-67) (SELBACH, 2013).

The 1965 Brazilian Forest Code was considered by many authors to be one of the most complete and advanced in the world, but reality shows that the law alone was not enough to guarantee real conservation on rural properties. Despite its importance, this law was the target of constant criticism and questioning and, as a rule, was never systematically observed by rural producers (GIEHL, 2007; DELALIBERA *et al.*, 2008; OKUYAMA *et al.*, 2012).

In practical terms, when the 1965 Forest Code was in force, there was no efficient mechanism for environmental organizations to demand compliance. Furthermore, this is not a problem

exclusively from disobedience. The efforts invested in communication were insufficient and many landowners were possibly unaware of a large part of their obligations.

The state of Rio de Janeiro is a good example of the failure to apply the old Forest Codes. In the Serrana, Noroeste and Vale do Paraiba regions, the expansion of coffee cultivation, followed by extensive livestock farming and, finally, conventional olive growing to replace the Atlantic Rainforest, contributed to the appearance of erosion, the reduction in the number of springs and the contamination of water resources. In addition, it is estimated that between 1985 and 2012 there was a decrease of 177,000 hectares of Atlantic Forest (FUNDAÇÂO SOS MATA ATLANTICA, 2014). The map below shows the location of the Atlantic Forest remnants in the state of Rio de Janeiro in 2013.

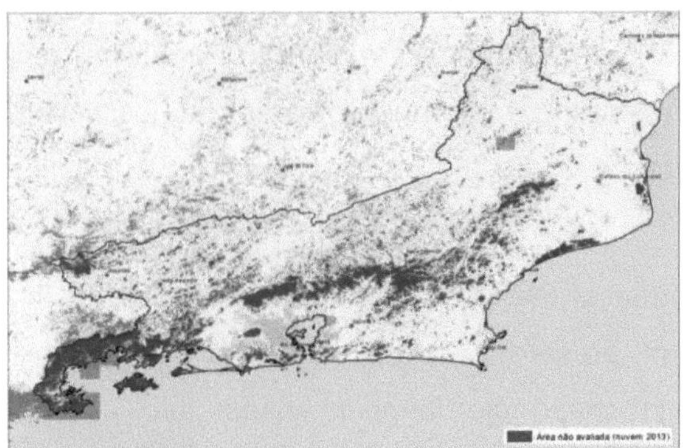

Figure 3: Map of the state of Rio de Janeiro with remnants of Atlantic forest highlighted in green. Source: Atlas of Atlantic Forest Remnants, 2012-2013.

As can be seen on the map, the state's remaining forests are concentrated in UCs and mountainous regions. Some authors, such as Tourinho (2005),

Kluck *et al.* (2011) and Okuyama *et al.* (2012), have highlighted the difficulties faced by rural producers in adapting to the environmental standards established in the 1965 Forest Code, confirming the inefficiency of the application of the 1964 law in the historical context.

The research carried out by Kluck *et al.* (2011) showed that there would be a negative socio-economic impact if the rules of the 1965 Forest Code were applied to banana farms in the municipality of Luis Alves in Santa Catarina. The results showed that with the implementation of the APPs, the banana farms analyzed would have a reduction in crop area of between 28% and 30%. The net income of the farms would be reduced by between 38% and 40% if the APPs were adapted. For the author, compliance with the APPs legislation would generate a very significant impact on the rural properties analyzed, to the point of making them unviable in their current form of production, and this issue becomes more critical the smaller the area of the property.

Okuyama *et al.* (2012) corroborate Kluck's (2011) position, based on the assumption that this legislation is inadequate, especially for family-based agriculture. For Tourinho (2005), environmental preservation cannot take precedence over the survival of the rural producer who makes a living from his property. The author states that there is a need to adapt the legislation, with incentives or feasible alternatives so that rural producers can comply with it.

It is worth pointing out that none of these analyses take into account the socio-environmental gains from preserving forest remnants, nor do they present alternatives that could shape the farming system into a more harmonious relationship with the protected areas. Of course, this doesn't detract from the authors' questions, but it does show that the issue is complex, highlighting the need to analyze other perspectives.

2.3 **The road to the new law**

Historically in Brazil, large landowners have always had the power to influence the state. Barcelos and Berriel (2009) discuss Getúlio Vargas' political pact with the traditional rural oligarchies at the end of the Old Republic as an example of the usual bargaining between landowners and public power. According to the authors, this strategy guaranteed the support of the big landowners for the federal government in exchange for not including the rights of rural workers in the bills that were being developed at the time.

Until the 1950s, agricultural work predominated in Brazil. In the second half of the 20th century, with the advent of the Green Revolution, a considerable portion of the countryside was unable to adapt to the new production patterns. At the same time, the country's economy was turning towards industry and these circumstances together caused a large part of the rural workforce to flee the countryside, increasing the power of the cities and generating a significant rural exodus. From then on, entities related to large landowners mobilized to combat this supposed "penalization of agriculture" with more organized participation in the political sphere (MENDONÇA, 2005).

In the 1980s, when the Forest Code reform agenda was already being discussed, rural bosses created an efficient strategy to push their agenda forward. The tactic, according to Pereira (2013), was to stifle the class struggle and construct a new discourse as if rural workers, businessmen and landowners had the same interests and needs. Throughout the Forest Code process, rural businessmen appropriated the discourse of defending small landowners and family farmers to defend their own interests. Since then, the so-called "ruralist caucus" has structured itself in Congress and currently consists of a supra-partisan group (the Parliamentary Agricultural Front), with representatives from various parties, from conservative to center-left, with representatives allied to the government and opponents from all regions of the country. Because of this characteristic, the ruralist caucus is able to defend its interests and is highly representative at national level (PEREIRA, 2013).

As Pereira (2013) points out, the sheer number of parliamentarians in the caucus is not in itself what makes it strong. The group's great asset is its ability to mobilize other parliamentarians not directly involved with the agenda. This characteristic has given the caucus some victories in votes of interest to them. Examples include the approval of the Biosafety Law[14] , the final report of the CPMI on Land[15] and the much-talked-about reform of the Forest Code.

From 2008 to 2012, the Forest Code was much discussed in the legislature, where there was also intense mobilization by opposing social movements, which even organized a movement known as: "Veta Dilma", in order to have the new law vetoed in its entirety. However, this did not have much effect. In May 2012, President Dilma Rousseff vetoed 12 (twelve) points of the law and proposed changes to 32 (thirty-two) other articles, but on May 25, 2012 the final text of the new Brazilian Forest Code was published through Law 12.651, modified by Law 12.727, sanctioned on October 17, 2012 for the regularization and adaptation of APPs and RLs on Brazilian rural properties (SELBACH, 2013).

According to Valera (2014), the approval of this law ignored the manifestations of scientific institutions such as the University of São Paulo (USP), the Luiz de Queiroz School of Agronomy (ESALQ) and the Brazilian Society for the Advancement of Science (SBPC), among others. Authors such as Valera (2014), Moura (2014) and Pereira (2013) believe that the

---

14Law 11.105/05. Approved by Congress and sanctioned with seven vetoes by the president in March 2005. The legislation regulates the planting and marketing of transgenic products.
15In November 2005, the Joint Parliamentary Commission of Inquiry (CPMI) rejected the opinion of the rapporteur, deputy Joâo Alfredo (PSOL/CE), and approved a separate vote by deputy Abelardo Lupion (DEM/PR) - a ruralist representative - which hindered the progress of agrarian reform policies and criminalized the struggle for land, materialized in the actions of the Landless Workers' Movement (MST) (PEREIRA, 2013).

reform of the Forest Code was not intended to improve the management of rural areas, forests and biodiversity, but rather to benefit the economic power of corporate agribusiness interests.

In order to promote a more specific investigation into the impacts of this new law on the study area, the following subchapter presents an analysis of the main changes to the new forestry code.

2.4 **The new Forest Code and its main changes**

Before presenting what the title of this subchapter proposes, it is important to emphasize that this research will only deal with aspects of the law that may have some social and/or economic impact on the activities of farmers in the Campinas MBH. Therefore, articles of the new Forest Code referring to biomes, cultural characteristics and geographical features other than those that occur in the study area will not be dealt with here.

For a better contextualization of the terms used hereafter, some of the basic definitions provided by the 2012 Forest Code will be presented. According to Article 3, the following definitions apply:

IV - consolidated rural area: area of rural property with anthropic occupation prior to July 22, 2008, with buildings, improvements or agroforestry activities, in the latter case allowing the adoption of the fallow regime;

V - small family rural property or possession: that which is exploited through the personal work of family farmers and rural family entrepreneurs, including settlements and land reform projects, and which complies with the provisions of art. 3° of Law no. 11.326, of July 24, 2006;

VI - alternative land use: replacing native vegetation and successor formations with other land cover, such as agricultural, industrial, energy generation and transmission, mining and transportation activities, urban settlements or other forms of human occupation;

VII - sustainable management: the management of natural vegetation in order to obtain economic, social and environmental benefits, respecting the mechanisms for sustaining the ecosystem being managed and considering, cumulatively or alternatively, the use of multiple timber or non-timber species, multiple flora products and by-products, as well as the use of other goods and services;

[...]

XVII - spring: a natural outcrop of groundwater that is perennial and gives rise to a watercourse;

XVIII - waterhole: natural outcrop of the water table, even if intermittent;

XIX - regular bed: the channel through which the waters of a watercourse flow regularly throughout the year;

[...]

XXV - wetlands: wetlands and terrestrial surfaces covered periodically by water, originally covered by forests or other forms of vegetation adapted to flooding.

Among the main changes brought in by Law 12.727 of 2012 are the following:

1 - According to article 15, APPs can be included in the percentage of preserved area required for the Legal Reserve. However, if the sum of the Legal Reserve and APP areas exceeds the required percentage, the producer cannot remove the excess vegetation;

2 - According to item I of article 4, in riparian forests, the width of rivers, which used to be measured by the largest bed reached during flooding, is now measured by the regular channel. However, this does not free producers to reduce their existing APPs. If the change in measurement reduces the size of the required APP, the surplus riparian forest can be used to compensate for Legal Reserve deficits on other properties.

3 - According to article 61-A and paragraph 1, agroforestry, ecotourism and rural tourism activities that already existed in APPs until July 22, 2008 are defined as consolidated areas and there is no longer any obligation to restore them, except for areas of riparian forest, which will have a different percentage of recomposition according to the number of fiscal modules of the property, and for springs on properties of any size, the preservation of only 15m will be required instead of the 30m that was required under the previous law.

4 - According to article 61, paragraphs 1, 2, 3 and 4, on properties smaller than 4 fiscal modules[16] with areas consolidated in APPs that have a river with

---

16The fiscal module corresponds to the minimum area needed for a rural property to be

a width of no more than 10 meters, the recomposition of the riparian forest must vary from 5 meters to a maximum of 20 meters, depending on the size of the property. Under previous legislation, the minimum recomposition required for any property was 30m, which could be as much as 200m depending on the river's flow. The table below illustrates this comparison.

APP class	Size (Rural Property)	General Rule APP (River Width)	APP Old Code	APP New Code	APP New Code Recomposition for Consolidated Areas (Art. 61)
Riverbank	Up to 1 MF	<10m	30m	30m	5m
	1 to 2 MF	10 a 50m	50m	50m	8m
	2 to 4 MF	50 to 200m	100m	100m	15m
	4 to 10 MF	200 to 500m	200m	200m	20m to 100m
	Above 10 MF	>600m	500m	500m	30m to 100m
Spring	All	All	50m	50m	15m

Table 1. Comparison of the minimum recovery bands required on riverbanks and springs in the 1965 and 2012 forest codes.

5 - According to article 66, in consolidated areas, regardless of adherence to the PRA, the Legal Reserve can be replanted through the natural regeneration of vegetation, by planting new trees.

(the use of up to 50% exotic species is permitted) or by offsetting (see item 8 below);

6 - According to article 67, on properties smaller than 4 fiscal modules it is no

economically viable. The size of the fiscal module for each municipality is set by Special Instructions (IE) issued by INCRA.

longer necessary to restore the Legal Reserves. The percentage of native vegetation existing on the property up to July 22, 2008 will apply, and there can be no deforestation of these areas after this date;

7 - According to paragraph 4 of article 64, rural landowners who choose to restore the Legal Reserve using interspersed planting of exotic species will have the right to exploit it economically.

8 - According to items I, II, III and IV of paragraph 5 of article 66, environmental compensation can take place outside the property through the purchase of an Environmental Reserve Quota (CRA), lease, donation to the public authorities of an area within a publicly owned conservation unit pending land regularization, or registration of an equivalent area in the same biome (BRASIL, 2012).

The rule on consolidated rural areas, conceptualized in item IV of Article 3, is one of the points most criticized by jurists and environmentalists. According to the Federal Senate's official website, jurist Paulo Affonso Leme Machado, who holds a master's degree in environmental law from the University of Strasbourg (France), believes that   the application of the concept of "consolidated rural area" encourages illegality and encourages the practice of environmentally disrespectful behavior, which is a form of "amnesty" without using that name. The rural caucus, for its part, defends this measure by emphasizing their commitment to respecting situations that have been consolidated over time, especially for family farmers (SENADO FEDERAL, 2015).

When analyzing the law, it can be seen that small rural properties actually receive greater protection from the rule of consolidated areas, because, from paragraphs 1 to 5 of article 61-A, the law establishes exceptions in which the smaller the property, the smaller the preserved area required. Below are the paragraphs to detail the situation:

Art. 61-A. In Permanent Preservation Areas, the continuation of agroforestry, ecotourism and rural tourism activities is exclusively authorized in rural areas consolidated until July 22, 2008 (Included by Law No. 12.727, of 2012).

§ 1o For rural properties with an area of up to 1 (one) fiscal module that have areas consolidated in Permanent Preservation Areas along natural watercourses, it will be mandatory to restore the respective marginal strips by 5 (five) meters, counted from the edge of the regular bed, regardless of the width of the watercourse (Included by Law No. 12.727, of 2012).

§ 2o For rural properties with an area of more than 1 (one) fiscal module and up to 2 (two) fiscal modules that have areas consolidated in Permanent Preservation Areas along natural watercourses, it will be mandatory to restore the respective marginal strips by 8 (eight) meters, counted from the edge of the channel of the regular bed, regardless of the width of the watercourse (Included by Law No. 12.727, of 2012).

§ 3o For rural properties with an area of more than 2 (two) fiscal units and up to 4 (four) fiscal units that have areas consolidated in Permanent Preservation Areas along natural watercourses, it will be mandatory to restore the respective marginal strips by 15 (fifteen) meters, counting from the edge of the channel of the regular bed, regardless of the width of the watercourse (Included by Law No. 12.727, of 2012).

§ 5o In the case of consolidated rural areas in Areas of

Permanent Preservation In the surroundings of springs and perennial waterholes, the maintenance of agroforestry, ecotourism or rural tourism activities will be allowed, with the recomposition of a minimum radius of 15 (fifteen) meters being obligatory.

Art. 61-B. Owners and possessors of rural properties who, on July 22, 2008, held up to 10 (ten) fiscal units and were carrying out agroforestry activities in areas consolidated in Permanent Preservation Areas are guaranteed that the requirement for recomposition, under the terms of this Law, when all the Permanent Preservation Areas on the property are added together, will not exceed:

I - 10% (ten percent) of the total area of the property, for rural properties with an area of up to 2 (two) fiscal modules; (Included by Law No. 12.727, of 2012).

II - 20% (twenty percent) of the total area of the property, for rural properties with an area greater than 2 (two) and up to 4 (four) fiscal modules (Included by Law No. 12.727, of 2012).

However, it is important to note that Article 61-A only establishes these exceptions for springs and watercourses. Features of great ecological importance such as hilltops and steep areas were not covered. In these cases, the landowner, who was already carrying out agricultural and forestry activities on other types of APPs than those mentioned in article 61, will not have the legal commitment to recover them, with the exception of landowners who until July 22, 2008 already had more than 10 (ten) fiscal modules. These will have to recover a larger area in accordance with article 61-B.

Hilltops are considered to be recharge zones, as they are essential for supplying the water table. They must be maintained with vegetation, as their function can be impaired if the soil is sealed as a result of compaction. Failure to preserve forests on hilltops and steep areas means that rainwater runs off the surface, causing erosion, siltation of rivers and the gradual nutritional depletion of the soil (MARTINS, 2001).

Another controversial point is the calculation of Legal Reserves in APPs. For producers who do not fall under the rule of consolidated areas, it will be possible to use the property's APP reserves to count the Legal Reserves, however, it may be that the rule of consolidated areas already covers a large part of the producers of MBH Campinas, since it is estimated that many of the agricultural areas were already consolidated before July 22, 2008.

Compensation is also a much criticized point. To this end, the following mechanisms could be used: lease by means of an environmental easement, outside the hydrographic basin and the state where the property is located, as long as it is in the same Biome; donation to the Government of an area located inside a Conservation Unit, pending land regularization or contribution to a Public Fund for this purpose; and the infamous acquisition of an Environmental Reserve Quota (CRA) - a title that represents native vegetation under an environmental easement, a Private Natural Heritage Reserve (RPPN) or a Voluntarily Established Legal Reserve on vegetation that exceeds the percentages established by law.

As it is no longer compulsory for properties smaller than four fiscal modules to recompose Legal Reserves, these properties can use all the natural vegetation outside the Permanent Protection Area (APP) as CRAs. As Barros and Barcelos (2016) point out, "This generates a quota for an area that has already been protected, i.e. there is no opportunity cost. It's an area that can't be touched". According to Victor Ranieri, a PhD professor at the University of São Paulo (USP):

"The quota has advantages from an economic point of view. It allows you to offset your Legal Reserve in areas where land is cheaper. But from an environmental point of view, this compensation is not always equivalent."

This analysis suggests that the application of CRAs in this way is out of step with the objective of protecting vegetation.

2.5 **The Rural Environmental Registry (CAR) and the Environmental Regularization Program (PRA)**

The new Forest Code, together with Decrees 7.830 of October 17, 2012 and 8.235 of May 5, 2014, made it possible to create the Rural Environmental Registry (CAR) and the Environmental Regularization Program (PRA) to guide and encourage producers to bring their properties into line with the environment.

The Rural Environmental Registry (CAR) is a nationwide electronic public registry, compulsory for all rural properties, with the aim of integrating environmental information on rural properties and possessions, forming a database for environmental and economic control, monitoring and planning. The application of this mechanism involves satellite georeferencing techniques that will make it possible to monitor the physical processes of land use and vegetation cover in order to support legislative and regulatory bodies in these tasks (BRASIL, 2012).

In addition, decree 8.235 instituted the Environmental Regularization Program (PRA), which will be managed by the states and the Federal District and consists of a plan to regularize the APPs and RLs with deadlines and targets that rural producers must meet according to the situation of their property in relation to the rules of the New Forest Code. For unregulated producers who do not adhere to the PRA, all fines for non-compliance with the old Forest Code will apply again (BRASIL, 2014).

The CAR decree was published on October 17, 2012, but the system was only implemented in May 2014, and initially a deadline of one year was set for landowners to register and begin the environmental regularization of their properties, but this date has been postponed several times due to the difficulty of making the registration process operational, since many producers do not have access to the internet and are not trained to fill in all the required information. According to an official statement from the Brazilian Confederation of Agriculture and Livestock:

"Despite being declaratory, the system has proved to be moderately complex to fill in, causing difficulties in the accurate provision of information and consequent delays in deliveries. If the CAR has not yet fulfilled its role, it is necessary to extend the deadline for registering rural properties as permitted by law," (official statement from the Confederation of Agriculture and Livestock of Brazil, 2015).

In some states and municipalities there are public institutions that help producers to register free of charge. In the state of Rio de Janeiro, INEA is responsible for this. However, the institution's services have proved to be insufficient and many producers have opted to register with private contractors.

The obligation to implement the CAR will be guaranteed by the restrictions imposed on landowners who fail to register on time. They will not have access to rural credit or to the benefits provided by the PRA, and their properties will be legally restricted, making it impossible to divide and sell them. It is important to remember that this linking of environmental compliance to rural credit[17] , was the major catalyst for mobilizing ruralists to adapt the Forest Code to their own interests. The CAR will also be a condition for landowners to register their property's "surplus protected areas" as CRAs. These areas could be the target of financial speculation through quota transactions, as discussed in the previous section (INESC, 2012).

17Originating from resolutions published by the Central Bank of Brazil.

Capitulo 3

Study area

3.1 **The watershed as a unit of planning and intervention**

From a geographical point of view, hydrographic basins can be understood as extensive rainfall catchment areas delimited by slopes (watershed), with a set of sub-basins with networks of drainage channels that flow into the sea. Each hydrographic basin interconnects with another of a higher hierarchical order, constituting, in relation to the latter, a sub-basin. Therefore, the terms river basin and sub-basin are relative (TEODORO *et al.,* 2007).

Micro-basin, in turn, is a term that came into widespread use with the advent of sustainable rural development programs supported by the World Bank. Currently, the term microbasin is a widespread nomenclature in Brazil that has been embraced by many authors such as Faustino (1996), Mosca (2003), Leonardo (2003), Cecilio and Reis (2006) and others. Some authors, such as Santana (2003), consider micro-basin to be an empirical term and suggest that the term sub-basin is more appropriate. However, we have opted to keep the term micro-basin in this work because current public policies that work with the concept of micro-basins have a great influence on our object of study.

The concept considered in this research considers the micro-watershed as a planning unit, corresponding to the level of verification, mediation, monitoring and intervention *in loco* where the relationship between human populations and the other biotic and abiotic components comprised by the surrounding slopes that form the drainage network that supplies these populations can be observed. Therefore, the concept adopted here is based on a socio-environmental perspective (RODRIGUES *et al.,* 2013; MOSCA, 2003;

LEONARDO, 2003). The figure below illustrates how a micro-watershed works.

Figure 4: Dynamics of a micro-watershed.

Source: Geoconceicao.

According to Silva (1994), in Brazil, integrated management in micro-watersheds began to be worked on in the first half of the 1980s in pilot experiments carried out in Paranâ. Currently, actions of this nature have been gaining prominence in public initiatives carried out in various Brazilian states with the support of important international financial institutions such as the World Bank.

The division of the municipality of Sumidouro into micro-watersheds (MBH) was carried out in 2010 by the Municipal Council for Sustainable Rural Development (CMDRS) in accordance with the guidelines established by the State Secretariat for Agriculture, Livestock, Fisheries and Supply (SEAPEC). The methodology employed aims to develop an action strategy that uses the MBH as a planning and intervention unit, directly involving the communities living in this geographic space.

This initiative is part of the Program for Sustainable Rural Development in Hydrographic Microbasins, also known as the Rio Rural program, which applies funds from the World Bank to actions for environmental preservation and income generation on rural properties.

Operated by Emater-Rio in the municipalities, Rio Rural was SEAPEC's main program for a long time. In MBH Campinas, between 2015 and 2017, 26 meetings were held to raise awareness and plan actions with family farmers. During this period, around 3,200 sub-projects were drawn up in the Individual Development Plans (PIDs), serving 786 producers in Sumidouro. MBH Campinas received the most funding, accounting for around 40% of total investments in the municipality. Among the main sustainable sub-projects developed are: green manure, direct planting/minimum cultivation, acquisition of irrigation equipment (drip irrigation), individual sanitation, protection of springs and protection of recharge areas.

However, despite being very well structured in SEAPEC's official documents, the program demanded quantitative results that conflicted with the qualitative development of the work. In addition, the delay in releasing the funds, coupled with the immediacy of the producers, ended up diluting a large part of the effort allocated to the sensitization process. In this way, much of what was encouraged was lost over time due to the lack of systematic follow-up, which is justified by the high volume of projects required and the low availability of human resources.

The legacy left behind was protected springs and recharge areas, a few fragments of well-recovered riparian forest[18] and a few pits and biodigesters installed. However, there has been no progress in terms of associations and the perpetuation of the agro-ecological practices encouraged.

3.2 **Social and environmental aspects of Sumidouro**

The municipality of Sumidouro has the highest percentage of rural population among the ninety-two municipalities in the state of Rio de Janeiro. Currently, 63.42% of its population lives in rural areas and has agriculture as its main

---

18 This resistance to restoring riparian forests is discussed in greater depth on pages 93 to 96.

source of income (IBGE, 2010). This rural predominance is an atypical reality within the state, given the growing emptying of rural areas and the consequent reduction in the importance of agriculture observed in recent years (ALENTEJANO, 2010). As can be seen in the following graph, there was a very sharp reduction in the population employed in agriculture in the state of Rio de Janeiro from 1985 onwards.

Graph 1. Comparison of the results of the structural data on the population employed in agriculture in the Agricultural Censuses - State of Rio de Janeiro - 1970/2006

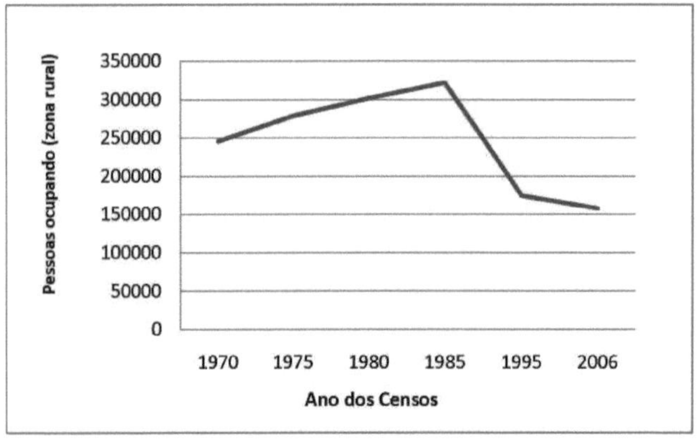

Fonte: IBGE, Censo Agropecuário 1970/2006.

Alentejano (2010) points out that this reduction was much greater than that observed in the country between 1985 and 2006. For the author, this is due to the decline of sugar cane, the abandonment of a large part of Rio de Janeiro's rural establishments and the cattle-raising of the remainder due to the phenomena of urbanization and the Green Revolution already discussed earlier in this research.

According to the 2010 demographic census, the rural population of the state of Rio de Janeiro is just 3.3% and agriculture has the smallest share of the state's Gross Domestic Product (GDP), accounting for just 0.6%. For these reasons, little is said about the importance of Rio de Janeiro's agriculture. However, it is well known that agriculture in the state of Rio de Janeiro "exists, resists and feeds". Even today, there are more than half a million people living in the countryside in the state of Rio de Janeiro and Sumidouro is a very representative municipality within this context (IBGE, 2010).

The municipality of Sumidouro is located in the mountainous region of the state of Rio de Janeiro. It has around 14,900 inhabitants and a geographical area of 396 km^2 (IBGE, 2010). The map below shows the municipality's location in the state.

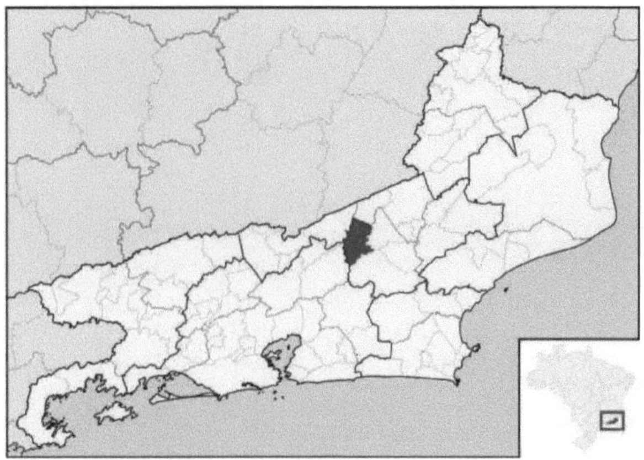

Figure 5 - Location of the municipality of Sumidouro in the state of Rio de Janeiro. Source: EMATER-RIO Annual Activity Report 2013.

The production of vegetables, legumes and fruit is favored by the geographical aspect that places the municipality between average altitudes of 400 meters and average temperatures of 27° C, in the region closest to the Paraiba Valley, and 900 meters and an average temperature of 22° C in the region bordering Nova Friburgo and Teresópolis (STOTZ, 2012). Sumidouro, together with the municipalities of Bom Jardim, Nova Friburgo, Teresópolis (Serra Fluminense) and Paty do Alferes (South Central Region), is the second main olive-growing center in the state, responsible for supplying the Metropolitan region, behind a few municipalities in the Northwest region (IBGE, 2015; PEREZ & MOREIRA, 2007). The following is a topographical map of Sumidouro, showing its division by watersheds.

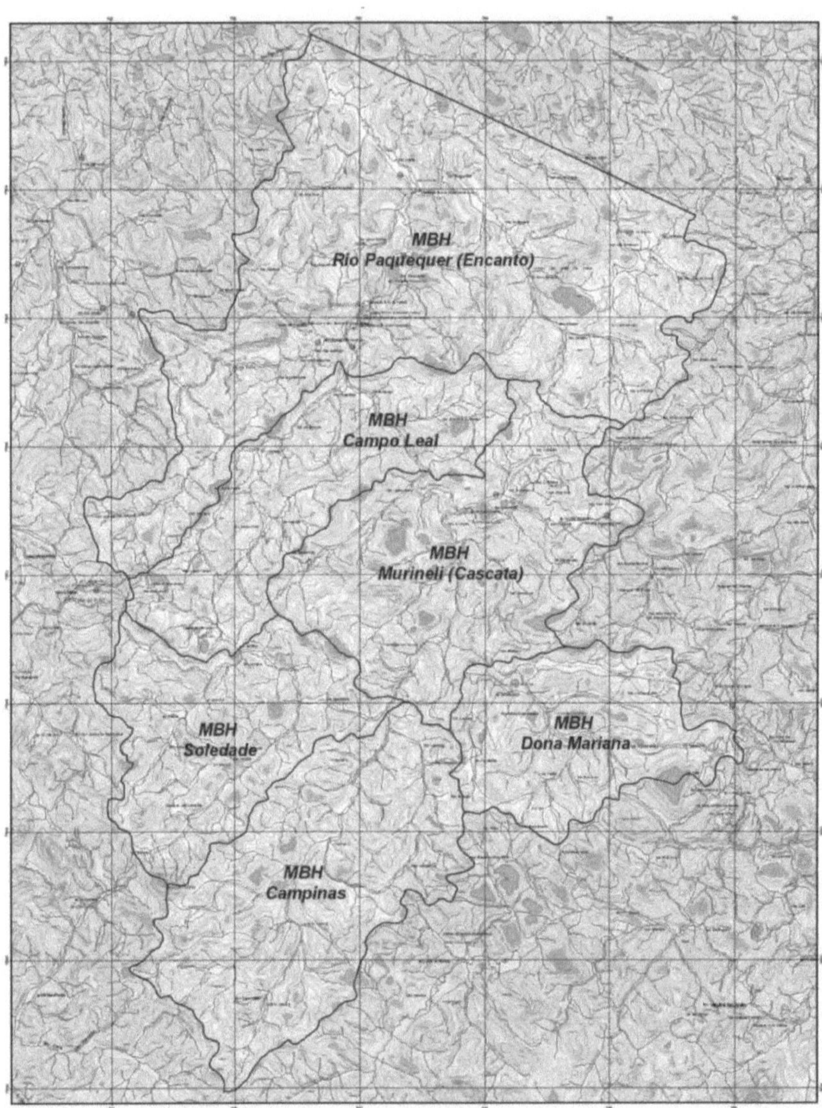

Figure 6 - Topographical map of Sumidouro with the division of the municipality by watersheds. Source: EMATER-RIO Annual Activity Report 2014.

3.3 **The Campinas watershed**

Of the five watersheds created by the CMDRS in Sumidouro, MBH Campinas is the one with the largest number of family farmers and the most representative in terms of agricultural production in the municipality. Located at high altitude, popularly known as the "cold land", MBH Campinas borders the municipalities of Teresópolis and Nova Friburgo in the southwest and southeast respectively.

The perimeter defined by the CMDRS as the boundary of MBH Campinas corresponds to the district of the same name. Because of this equivalence, it is possible to use the official data for the district to make inferences about the Campinas MBH. This is permissible because, coincidentally, the towns belonging to the Campinas district together form the geographic feature that characterizes a watershed (or sub-basin). The water that rises in each of the towns (Santo André, Sâo Bento, Vale dos Pinheiros, Aguas Claras and Caramandu) converges into the same main river, which receives various tributaries along the way and culminates in the Conde d'Eu waterfall, which flows into the Paquequer river. Below is a map showing the locations in the watershed and its hydrography.

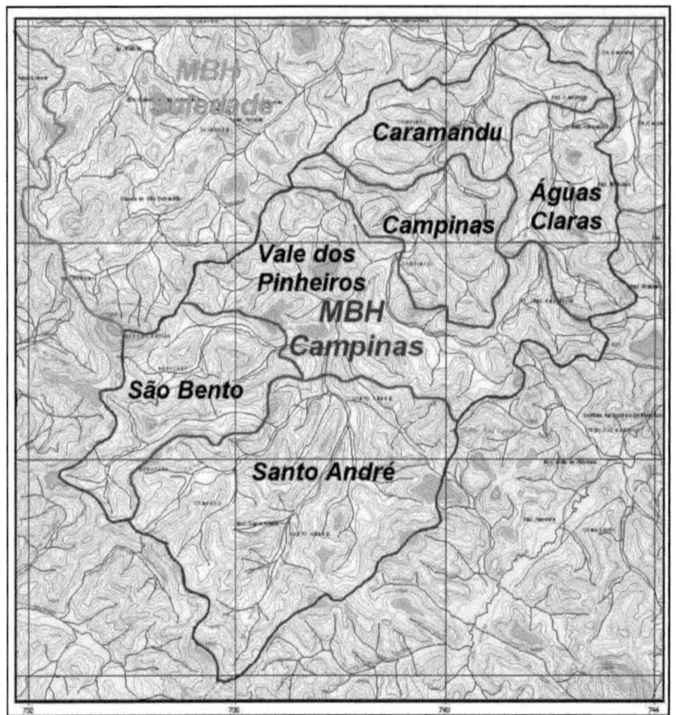

Figure 7. Division by locality of the Campinas Hydrographic Microbasin. Source: Emater-Rio collection.

According to IBGE data, the Campinas district has 3105 people living in rural households scattered around the towns. It is estimated that MBH has 1035 rural families. The district has a small commercial center where 617 people live. Therefore, the percentage of the rural population is 83.4% while the urban population is only 16.6% (IBGE, 2015).

The main activity of the rural population of the MBH is growing vegetables, with the greatest emphasis on the cultivation of box plants[19] and in second place leafy vegetables[20] . However, there is a small commercial center in the watershed where there is a growing expansion of underwear manufacturers, which generally use the labor of women, many of whom have left rural

19Boxes: organic produce sold in boxes. Examples: tomatoes, peppers, zucchinis, jelly beans, eggplants and cucumbers.
20 Vegetables in which the edible parts are the leaves, stems or inflorescences. Examples: lettuce, broccoli, cabbage, coriander, parsley.

44

activity, to supplement family income. There are also decentralized garment factories scattered around the localities and close to the cultivation areas. In these cases, families carry out both activities, in some cases with part of the workforce hired.

Teixeira (1998) already pointed to this phenomenon in the mountain region at the end of the 1990s. The author sees this process as a response to the impacts of agricultural modernization, with family pluriactivity being a strategy for social reproduction, since the internalization of industrial production and the flexibilization of the production process are seen as alternatives for supplementing income and even for families to remain in the countryside.

Turning to agriculture, in the Campinas MBH there are a considerable number of families who own small rural properties, the vast majority of which are smaller than one fiscal module (26ha). Based on the Participatory Rural Diagnosis (PRD) of the watershed and data from INCRA, it is estimated that there are 552 rural properties in Campinas. However, some families work in partnership on properties owned by third parties. In this case, the owner pays the production costs and keeps 40, 50 or 60% of the profits, while the partner keeps the rest.

These arrangements are difficult to characterize because they are very dynamic. Often, even landowners with productive areas work in partnership with a relative or neighbor at certain times of the year. In these cases, the partner families work alongside the owner family without any apparent employer-employee relationship. In many cases, there is no socio-economic differentiation between owners and partners. It is important to note that this arrangement does not necessarily de-characterize the landowner's activity as family farming. However, the partnership system can also occur due to three other distinct situations: the existence of farmers who do not own rural property; the existence of landowners whose area available for cultivation has become insufficient; the landowner is decapitalized, having neither the money nor the credit to finance farming.

Although the agriculture practiced in MBH Campinas is generally characterized as family farming, the production method in the region is based on capitalized activity and is highly dependent on external inputs. Family work and soil exploitation take place intensively, with daily activities involving weeding, fertilizing, pesticide application, irrigation and harvesting, as well as the tasks inherent in marketing, such as selecting, packaging, distributing and selling products (EGGER, 2010). The number of inputs in the production chain is enormous. Farmers buy limestone, fertilizers, a wide variety of herbicides, insecticides and fungicides, which makes the cost of production very expensive while favoring a gigantic market made up of half a dozen multinational manufacturers and distributors.

The municipality of Sumidouro is among the biggest consumers of pesticides in the state. Between 2009 and 2013, 120 cases of pesticide poisoning were recorded in the Notifiable Diseases Information System (SINAM), and the estimate of underreporting is that for each case there are 50 similar unreported cases (FUNDAÇÂO OSWALDO CRUZ, 2016).

In a survey carried out in Caramandu, a town belonging to MBH Campinas, it was found that none of the farmers interviewed used Personal Protective Equipment (PPE) properly when applying pesticides. In the same survey, there was also little evidence of the use of natural or alternative methods to control pests and diseases. Many of those interviewed showed that they were aware of the negative implications that pesticides can have on the environment and human health (OLIVEIRA, 2014). However, the production logic they are part of puts them in an immediate situation, in which any care that could "delay" or in any way "hinder" their work is met with resistance.

Stotz (2012) in his article - The limits of conventional agriculture and the reasons for its persistence: a study of the case of Sumidouro, RJ - points out that we must bear in mind that in so-called "family farming" the conventional agricultural system was imposed through subsidized credit and public technical assistance. Currently, Sumidouro is one of the municipalities in the state of Rio de Janeiro with the highest number of PRONAF operations for funding and investment. According to data from Emater-Rio, more than 30% of the municipality's credit operations come from MBH Campinas. Producers depend on credit to gain access to the means of production, since productive autonomy has been lost throughout history. Nowadays, few producers produce their own manure, use their own seeds or grow food for their own subsistence (OLIVEIRA, 2014).

The terrain of MBH Campinas is highly rugged. Many crops are grown in fragile areas, mainly on steep slopes and near streams and springs. Some rural producers even have their entire property with their homes, improvements and crops within APPs. Because it is a year-round crop, it requires permanent dedication. Under favorable soil and climate conditions, it is possible to grow three tomato crops or six lettuce crops in the same area (STOTZ, 2012). This intensive practice is very common and causes soil quality to degrade over the years. As a result, the process of deforestation and the spread of crops to new areas is still a reality.

Figure 11 shows a satellite photograph of the MBH Campinas, where it can be seen that in the most intensively cultivated areas (Campinas, Vale dos Pinheiros, Aguas Claras, Caramandu) the forest remnants are much more sparse and sparse, compared to the less intensively cultivated regions (Santo André and Sâo Bento). To understand the map, just look at the areas where dark gray predominates. These are the areas with the highest density of forest fragments.

Figure 8. Satellite photograph of MBH Campinas showing the proximity of the forest remnants to the lower production sites (Sâo Bento and Santo André). Source: Google Earth.

Under the old Forest Code, due to the lack of enforcement, many rural producers deforested Permanent Preservation Areas (APP) and did not set up Legal Reserves (RL). According to data from the NGO SOS Mata Atlàntica (2009), it is estimated that the remaining cover of Atlantic Forest in the municipality of Sumidouro is 18%, comprising 7,311 hectares out of a total area of 39,551 hectares. It should be borne in mind that the scale used does not differentiate between forest and capoeira, which suggests that deforestation may be even greater.

Capitulo 4

Methodology

4.1 Application of questionnaires

Initial data collection was carried out using the Survey method (FREITAS, *et al., 2000)* with the application of structured questionnaires using a sample of 15% of the total estimated number of rural properties in MBH Campinas.

Before going any further, it is important to clarify that the IBGE identifies, for statistical purposes, the basic unit of the rural environment as the agricultural establishment, understood as all land of continuous area, regardless of size or location (urban or rural), made up of one or more plots, subordinated to a single producer (or family unit), where agricultural exploitation takes place. INCRA, in its National Rural Registry System (Law 5.868/72), uses the rural property as the basic unit, which is understood as the continuous area that is or can be used for agricultural, livestock, plant extraction, forestry or agro-industrial exploitation, regardless of its location (urban or rural) (NEUMANN and DIESEL, 2006).

It should be noted that the concepts of rural establishment and rural property are not equivalent, since there can be several rural establishments on one rural property, as this includes fractions caused by leasing, lending, etc. In turn, the same establishment can be made up of several rural properties, as is common in the case of sugar and alcohol mills. As the Rural Environmental Registry monitors rural properties, I chose to use this unit as the basis for my data collection.

According to Sumidouro City Hall, based on INCRA data, the municipality has around 1680 rural properties. Based on the Participatory Rural Diagnosis of the Campinas Hydrographic Microbasin (2015), it is estimated that there are

552 rural properties in the Campinas MBH. It was therefore necessary to administer 83 questionnaires to cover the stipulated sample percentage.

Due to the peculiarities of the populations of the different localities that make up the MBH Campinas, it was decided to work with non-random samples, i.e. the target audience was chosen by establishing proportional quotas relating to the geographical distribution of producers by locality (CARNEVALLI & MIGUEL, 2001). Thus, localities with a smaller number of producers and lower production intensity, such as Santo André and Sâo Bento, together accounted for approximately 17% of the total number of questionnaires. Intermediate locations such as Caramandu and Aguas Claras together received approximately 32% of the questionnaires and in Campinas and Vale dos Pinheiros, which are the most populous locations with the highest production, approximately 51% of the total questionnaires planned for MBH were applied. In this way, it was possible to represent the diversity of the MBH more faithfully. Table 1 below shows the distribution of the questionnaires applied by location.

Table 1. Distribution of questionnaires by location

Location	Questionnaires applied
Vale dos Pinheiros	21
Campinas	21
Caramandu	14
Aguas Claras	13
Saint André	8
Sâo Bento	6
Total	83

According to the classifications of Boyd & Wetfall (1964), the tool used in this research consisted of an "undisguised structured questionnaire", in which the respondent knows what the objective of the research is, and the document is standardized and has closed questions. The aforementioned document had questions on socio-economic data, the use of water resources, sanitation and protected areas on rural properties. The questions led to answers limited to numerical, ordinal, nominal and interval variables. A simple vocabulary was used in the wording of the questions, adapting the language so that the questions were easily understood by the interviewees, avoiding ambiguous interpretations.

Part of the questionnaires used were distributed at the Rio Rural program meetings, preceded by a brief explanation of how to fill them in. Producers with poor reading and interpretation skills were instructed to ask for help from other people, including me, who was on hand. On these occasions, the producers were also given a free and informed consent form which explained the objectives of the research and made it clear that the names of the participating producers would not be disclosed and that this procedure would not lead to any burden or punishment for whatever the content of the answers.

After this preliminary stage, with the first questionnaires in hand, it was realized that it wasn't possible to meet the quotas set per location due to the absence of producers from the more distant regions of MBH at the meetings, and the large number of questionnaires filled out incompletely. As a result, around 70% of the questionnaires had to be applied through interviews conducted by me during visits to the properties.

It was decided to use in the survey only the questionnaires answered by owners or heirs who work on the site, as they tend to have greater knowledge of the characteristics of their property, as well as being responsible for the environmental suitability of their rural property. With the questionnaires in hand, the data was statistically analyzed by grouping the patterns found. The questionnaire model used can be found in Appendix A.

4.2 Focus group interview

Morgan (1997) defines focus groups as a qualitative research technique, derived from group interviews, which collects information through interactions between participants. The application of this method was intended to meet the subjective needs of this research. The focus group interview took place in Caramandu in November 2015 with a heterogeneous group of farmers including men, women, young and old from different parts of the watershed. There were nine participants in total, all of whom were landowners or heirs. The first step in using this technique was to draw up a debate script with the topics to be discussed in the groups. This script had just four key questions broken down into subtopics through which the discussions were conducted.

To carry out the method, in accordance with Dias (2000) and Neto, Moreira and Sucena (2002), I invited two researchers I trusted to help me in the roles of observer, reporter and recording operator. We divided up the tasks as follows:

During the focus group:

- Leandro (Me) - moderator

- Kênia (pedagogue at the Municipal Department of Education and Culture) - observer and rapporteur

- Marcelo (professor of biological sciences and PPGEAS student) - observer, rapporteur and recording operator.

After the focus group:

- Leandro (Me) - transcriber and typist

Here is a brief description of each of these functions according to Neto, Moreira and Sucena (2002):

Mediator: The key function of the technique. The mediator is responsible for initiating, motivating, developing and concluding the debates, and is the only person who should intervene and can interact with the participants. The quality of the data and information gathered in the focus group is closely linked to its performance, which translates into (a) fostering the integration of the participants; (b) ensuring equal opportunities for all; (c) controlling the speaking time of each participant and the duration of the FG; (d) encouraging and/or cooling down debates; (e) valuing the diversity of opinions; (f) respecting the participants' way of speaking; and (g) refraining from influencing and opinion-forming positions.

Rapporteur: His task is to record the speeches, naming them, associating them with the reasons that prompted them and emphasizing the ideas they contain. They should also record the participants' non-verbal language, such as tones of voice, facial expressions and gestures. The material produced doesn't have to be a verbatim transcript of the speeches - this task falls to other functions - but rather a list of attitudes, ideas and points of view that will inform subsequent analysis.

Observer: This function aims to analyze and evaluate the process of conducting the focus group, paying attention to the participants in isolation and in their relationship with the Mediator, Rapporteur and Recording Operator. Your notes should aim to constantly improve the quality of the work and overcome the problems and difficulties faced, taking as a starting point (a) whether each participant felt at ease in front of the professionals; (b) whether there was integration between the participants; (c) whether they correctly understood the purpose of the research and (d) the way in which the roles of Mediator, Rapporteur and Recording Operator were carried out.

Recording Operator: A function designed to record debates in their entirety - depending on the equipment available.

Transcriber: Generally seen as an accessory or even a subordinate, this function is also important because, if not carried out properly, it can alter the participants' speech, which will cause serious damage or even make it impossible to correctly analyze the information obtained. The transcription must be as faithful as possible, with no interpretations, "text cleaning" or "copying" of the statements. All language errors, as well as pauses in the dialogues, should be kept and noted so that the analysis is as good as possible.

Typist: Like the previous one, this function is often mistakenly underestimated. Their task is to transpose all the data, handwritten or not, systematized, coded or recorded into a computer program, using the most appropriate software that provides the desired result.

The focus group interview took place as follows: we received the invited farmers at the Joâo Marchito school in Caramandu at the stipulated time and set aside the first twenty minutes for a chat and a snack in order to create a more friendly atmosphere. I then introduced the researchers who were helping me, the aims of the research and explained the focus group methodology, making it clear that the interview would be filmed and that the data obtained could be used in academic publications. I then asked the participants to sign an authorization form for the use of their images and testimonies and a free and informed consent form.

The chairs were positioned in a semicircle and with the help of a projector, we began the focus group with the first of the key questions. Each question was read out by me and then opened up for free discussion by the group. As the topics came up in the speeches, I, as the mediator, would mark them on the script to indicate that they had already been addressed. During this process, I tried to make the transition from one key issue to another as subtle as possible, always trying to follow the natural course of the discussions. The simplest topics were tackled first, gradually moving on to the more complex ones. Throughout the process, discussions arose that were not foreseen in the original script, but as they fitted in with the proposed objectives, I chose to stimulate them and later incorporated them into the script in the form of sub-topics. The activity lasted two hours. The model of the script used can be found in Appendix B.

As for the recording equipment, we used an Olympus video and audio camera fitted with a 24mm wide-angle lens, positioned to film the environment in a frame in which it was possible to capture the image of all the participants without the need for constant manipulation, and a Motorola Moto G2 cell phone to capture the audio, positioned next to the producers who were furthest away from the camera. The audio recorded by the cell phone was not used afterwards, as the sound captured by the camera was satisfactory for understanding the speech of all the participants, regardless of their position in the room.

Secondly, I watched the videos and transcribed them. After typing up all the audio from the focus group, the material was analyzed using the Content Analysis method proposed by Bardin (1977).

Content analysis is a set of research techniques that can be applied to qualitative research in different fields and for different purposes, as long as it uses oral or written communication as its object of study. The interpretation of these communications is based on a quantitative systematization that consists of identifying and grouping the nuclei of meaning, and then analysing them based on their presence or frequency. For the organization of Content Analysis, on page 95, Bardin (1977) highlights three important stages to be completed: 1- pre-analysis; 2- exploration of the material and 3- treatment of the results, inference and interpretation (BARDIN, 1977).

Pre-analysis is the phase of organizing the material. At this point, everything I transcribed was the subject of a floating reading that allowed me to select what would be analyzed and formulate some hypotheses, also taking into account the observations made by the researchers who helped me carry out the focus group. During this stage, I allowed myself to be imbued with impressions that enabled me to draw up indicators that helped support my interpretation of the results.

Next, I explored the material, separating the nuclei of meaning, regrouping the recurrences around the axes analyzed, to finally proceed to the last stage, which was the treatment of the results, inference and interpretation, in which I interpreted the material obtained from the analysis according to the theoretical framework of this research.

Capitulo 5

Results and Discussion

5.1 **Analysis and discussion of questionnaire results**

The overall consolidated result is represented in the percentages obtained as shown in the tables and graphs presented from page 74 to page 82.

5.1.1 <u>Size of properties</u>

Graph 2. Size of the properties analyzed

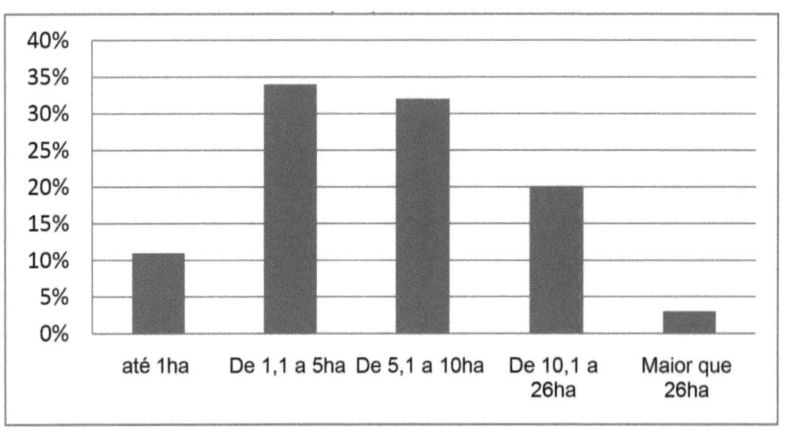

Average size of the properties analyzed: 7.5ha.

Bearing in mind that this study only counted landowners, the average size of the properties analyzed (7.5ha) was in line with expectations. The small size of the properties is due to the low levels of rural exodus that have occurred in Sumidouro and to hereditary division, where parents divided their properties for their descendants (EGGER, 2010).

It can also be seen that a very small number of the properties analyzed (3%) exceed the size of the municipality's fiscal module, which is 26ha. In other words, 97% of the landowners analyzed can benefit from the new Forest

Code's riparian forest preservation rule, which requires the preservation of only five meters on each bank. The 3% of landowners whose properties exceed one fiscal module do not reach two fiscal modules, so they will only be required to preserve eight meters of riparian forest. If these producers don't complete their CAR by the deadline, the rule of 30 meters of forest under the 1965 Forest Code will apply again, in addition to the legal restrictions and charges.

5.1.2 <u>Owners' education</u>

Graph 3. Overall consolidated result of the level of education of the owners interviewed

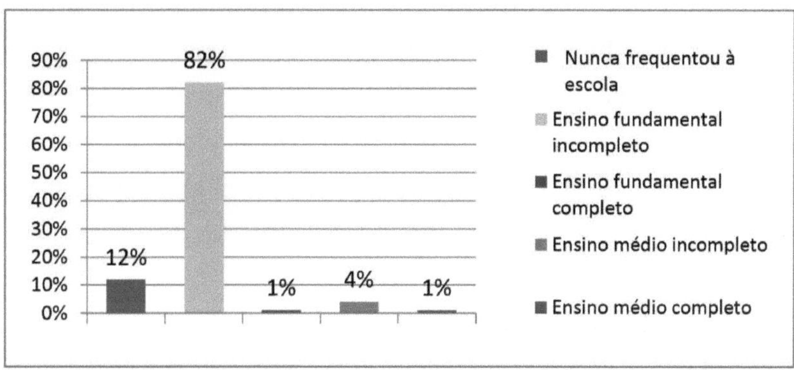

The results found for schooling in the Campinas MBH corroborate existing data from the survey carried out in the Caramandu locality, according to Oliveira (2014). From this analysis, it was possible to identify that low schooling is a problem that affects the watershed as a whole.

The 6% of landowners who had finished elementary school or continued their studies were all young people under the age of 30. Because this survey was focused on the landowner, a small number of young rural people were interviewed, as many of them work under a loan agreement on their parents', in-laws' or uncles' property. It is believed that if we had a larger number of young people in the survey, the level of schooling obtained might have been a little higher, since the generations from the 1990s onwards have found it easier to reconcile work in the countryside with their studies, due to the existence of schools scattered around the localities and public school transport.

However, it should also be borne in mind that, even today, schooling is still not highly valued by rural families. In this survey, among the 82% who had not completed elementary school, there were 5 young people under the age of 30, which represents 6% of the total. In the survey carried out in Caramandu in 2014, 31% of those interviewed were between 18 and 30 years old and even so, 94% of the total number of participants had not completed the first segment of elementary school (OLIVEIRA, 2014).

Most of the people who claimed to have incomplete primary education stopped their studies in the old fourth grade (now the fifth grade) because the age reached at this stage used to be considered sufficient for work, and also because the schools that offered the second segment of primary education onwards were fewer in   number and far between for producers from regions further into the MBH.

When comparing the national averages for schooling with those of Sumidouro, according to the 2010 IBGE Census, we notice that the municipality is well below, especially in the "no schooling" category, where the national average is 23.22% and Sumidouro's is 41.04%. The fact that the municipality is predominantly rural contributes greatly to this result. If we compare Sumidouro with other predominantly rural municipalities, we will most likely get equivalent figures, after all, low schooling in rural areas is a common problem throughout the country. According to a diagnosis carried out by the IBGE in 2012, the highest illiteracy rates in Brazil are in rural areas. This study found that the national rural illiteracy rate was 21.2% (IBGE, 2012).

It is the school's job to stimulate the critical awareness of its students.

According to the Law of Guidelines and Bases of National Education (LDB, Law No. 9.394, of December 20, 1996), in its article 35, item III, secondary education has the purpose, among others: "the improvement of the student as a human person, including ethical training and the development of intellectual autonomy and critical thinking". In this way, low levels of education can have an influence, not only on the low level of understanding of the ecological concepts that guide the biodynamic principles of agricultural production, but also on the critical ability to question the condition of exploitation to which producers are subjected when they produce in the conventional system. However, it is important to note that school is only one part of citizen education. The experience acquired throughout life can also contribute to critical education, and school doesn't always manage to fulfill its role in this respect.

5.1.3 Sanitation and use of water resources

Table 2. Consolidated general results on sanitation and the use of water resources in the MBH Campinas.

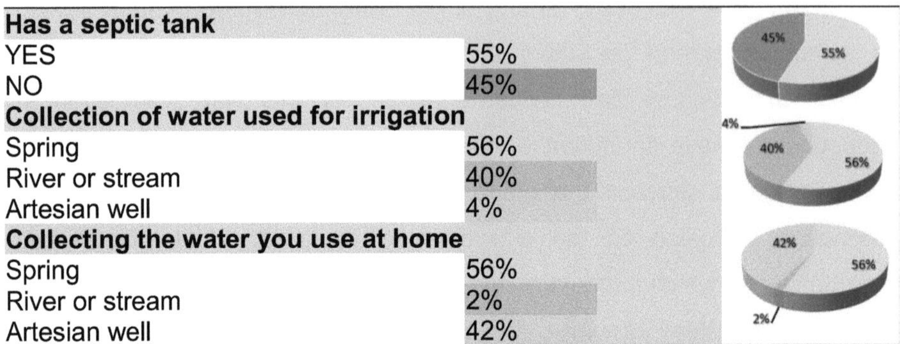

Has a septic tank		
YES	55%	
NO	45%	
Collection of water used for irrigation		
Spring	56%	
River or stream	40%	
Artesian well	4%	
Collecting the water you use at home		
Spring	56%	
River or stream	2%	
Artesian well	42%	

Firstly, it's important to note that this data was included in the survey because the lack of basic sanitation directly affects water quality and, likewise, the types of uses of water resources help to characterize the importance of having recharge areas, springs and riparian forests protected.

This analysis shows that almost half of the farmers interviewed do not have any type of domestic effluent treatment. This waste is channeled directly into watercourses and yet 40% of the farmers interviewed use streams and rivers to irrigate their crops, which can lead to the production of contaminated food that can have a harmful impact on consumer health.

This untreated sewage, to a lesser extent, can also affect the quality of water from springs and artesian wells depending on how close they are to the house, damaging the quality of the water used by the owner's own family.

It was also observed that springs are the most used sources of water for both irrigation and domestic use in the watershed (56% for both). Generally, producers make a reservoir to accumulate the water from this spring and pump it to the crops from there. Artesian wells also proved to be very important for domestic use.

5.1.4 <u>Situation of protected areas in MBH Campinas</u>

Table 3. General consolidated results of the protected areas situation in MBH Campinas.

Has woodland on the property (with more than 50% native species, except capoeiras and fallow areas)	
YES	76%
NO	24%
Size of forest on the property (for those who answered "YES" to the previous question)	
Less than 1ha	45%
From 1 to 3ha	41%
From 3.1 to 8ha	11%
Larger than 8.1ha	3%
Has springs on the property	
YES	72%
NO	28%
Number of springs on the property (for who answered "YES" to the previous question)	
1 spring	55%
2 spring	27%
3 springs	18%
It has forests around the springs (for those who answered "YES" to the question about springs)	
YES	47%
NO	53%
Is there a river or stream running through the property?	
YES	77%
NO	23%
Does it have woodland around rivers or streams (with more than 50% native species, except capoeiras and fallow areas) (for those who answered "YES" to the question about rivers and streams)	
YES	19%
NO	81%
Do you consider the area available for cultivation on your site to be sufficient?	
YES	78%
NO	22%

On the question "do you have forests on your property?", 76% of landowners claimed to have forests with more than 50% native species[21] , except for

---

21 It's important to note that when asking this question, the producers were explained the meaning of native forest.

capoeiras and fallow areas on their properties. This is a significant number, which shows that even though the new law acts more to maintain the remaining areas of forest than to encourage the restoration of forests in exploited areas, there are still a considerable number of forest fragments in MBH Campinas.

In most of the properties analyzed (45%), the conserved area does not exceed 1ha, which shows that perhaps the new law does not favor the occurrence of ecological corridors[22] linking the forests of the properties, since producers who fall under the rule of consolidated areas will not be obliged to recover green areas in addition to those that already exist on the property.

As for the existence of springs on the properties analyzed, 72% of the owners claimed to have them. As seen in Table 3, springs are the most widely used sources of water for both irrigation and domestic use in the watershed. However, 53% of the interviewees claimed that there was no forest around them. This item did not take into account the radius of the forest around the outcrop, or even whether it was native species or for economic use. Therefore, this 53% refers to properties that do not have any perennial species protecting the water resource. During the questionnaires, some producers commented that there used to be a larger number of springs on their properties, but that they had dried up.

With regard to riparian forests, the results obtained were more critical, as 81% of the producers claimed that they had no vegetation on the banks of the rivers. This can easily be seen when you walk around the watershed. There are many crops, buildings and pastures on the banks that can be seen along the road, but as there are many streams that pass within the properties, far from the roads, it was important to include this question in the questionnaire.

When asked, "Do you consider the area available for cultivation on your farm

---

22The aim of implementing an ecological corridor is to reduce fragmentation by maintaining or restoring landscape connectivity and facilitating the genetic flow between populations (MMA, 2015).

to be sufficient?", 78% of respondents said yes. This shows that although the properties are small, in general this proportion is still adequate for the type of cultivation practiced in the watershed. However, it is known that conventional cultivation, the way it has been practiced in the region, tends to exhaust the productive potential of the soil, which can lead to the search for new areas that are still fertile, as well as other factors, such as the search for increased production in order to maximize profits, or the progressive division of properties for reasons of inheritance.

The 22% who claimed that their land was insufficient for cultivation were not necessarily the ones with the smallest areas, as 8% of them had between 5.1 and 10ha. However, the amount of forested area on the property does not seem to be the main cause of this lack of arable land, as only 14% of the total properties analyzed had more than 6ha of woodland. This shows that the productive capacity of the family and the number of sharecroppers on the property are aspects that also have a direct influence on this issue.

5.2 Analysis and discussion of the results obtained in the focus group

In order to present the results and discussion of the focus group in a clear, objective and cohesive manner, the speeches presented here have been organized according to the central idea of the subjects and not necessarily in the order in which they occurred. Speeches by a participant that express the same ideas as those they had already expressed at a previous time and conversations that refer to subjects outside the scope of this research have been removed from the analysis. In order to preserve the identity of the participants in the writing of this chapter, their names have been replaced by Arabic numerals according to the order in which they were positioned in the room. The names of stores and people mentioned during the conversations have been replaced by X and Y.

5.2.1 <u>Knowledge of the New Forest Code</u>

Only participants 1, 2, 4 and 7 spoke about their own knowledge of the law, as can be seen in the statements below:

It's 20% of the land that has to be cleared of undergrowth, right? That's what the environmental registry is for (participant 4).

I've heard a bit about it. There are areas of the land that have to be cleared of undergrowth (participant 1).

You have to preserve 6 meters of the riverbank and the springs (participant 2).

And at the headwaters, right? (participant 7's speech complements participant 2's).

Participants 4 and 7 showed a certain insecurity when they ended their speech with the expression "right", but they still answered correctly, but without going into further detail. Participants 5 and 6 showed their agreement with participant 2 through body language by nodding their heads positively. However, participant 2 made a slight mistake when she stated that the required preservation margin is 6 meters. It is believed that participant 2 was referring to the owners of sites smaller than 1 fiscal module, which fall under the rule of consolidated areas (according to 98% of the 83 properties analyzed in this research). In this case, the area of riparian forest to be protected would be 5 meters instead of 6.

At no point were the terms Permanent Preservation Area (APP) and Legal Reserve (RL) mentioned. Nor was there any reference to the fact that the law had changed. However, even though there is little familiarity with the nomenclatures and some confusion about the measurements and other details relating to the New Forestry Code, in general it was possible to see that the participating producers have some knowledge of their obligations under the law.

5.2.2 <u>The importance of forests on rural properties</u>

According to the participants, it was possible to identify four central ideas regarding the importance of forest remnants on rural properties, as can be seen in table 4 below:

Table 4. The importance of forests on rural properties according to the farmers who took part in the focus group.

Forests are important for:	Opinion held by:
Human health	Participants 1 and 6
Regular water flow	Participants 1 and 2
Animal habitats	Participants 8 and 6
Agriculture	Participants 1, 4 and 5

The idea that forests are important for human health was firstly defended by participant 1, in a speech with little argument and a certain amount of sentimentality:

[...] I think the forest is very important for human health. You go into the forest and you feel good. It seems to change your system (participant 1).

Participant 6 introduces the idea of habitat for wildlife and adds at the end that forests are important for human respiration:

Species of birds and animals are extinct because deforest here and they flee there, deforest there and they flee further away. Not to mention our breathing. We see that plants are the lungs of the world (participant 6).

At this point, it's interesting to note that participant number 6's speech brings up a famous jargon - "forests are the lungs of the world" - an erroneous statement from an ecological point of view, since forests not only produce oxygen through photosynthesis, but also consume it through cellular respiration. The oxygen that is released into the atmosphere in the process of photosynthesis is mostly produced by marine phytoplankton. However, what we should consider in this statement is the realization that forests have an influence on air quality, directly affecting the health of living beings.

On the question of habitat for animals, participant 8 agrees with participant 6 and adds:

The forest also helps nature. The animals, too. Birds and other animals have to have their own place. With deforestation, the animals are disappearing. They're looking for another place (participant 8).

In the passage "the forest helps nature", we see a conceptual confusion that brings us back to Porto-Gonçalves (1996). In his work, the author points out that the concept of nature is socially constructed and changes according to place and culture. Here, the producer uses the term nature to refer only to animals. Later on, in the sentence: "[...] of course we're going to look for a way to benefit ourselves, but so that we don't harm nature, the manufacturer who couldn't make it[23] " (referring to agrochemicals), the producer uses the term nature again, but this time in a broader sense. In both situations, he sees himself as an individual who needs nature, but who doesn't identify himself as an integral part of it.

At another point, participants 4 and 5 introduce the idea that forests have beneficial effects on agriculture.

The bush doesn't get in the way, you just can't plant in the bush because of the shade. The bush is cold, it doesn't let many insects breed. You can see that in cold weather there are far fewer insects. The forest is good for that (participant 4).

Depending on where you are in the forest, the sun can shine more in the morning or in the afternoon. If it doesn't, the crop won't grow, but if it's only in the morning or afternoon, it's great, because it's cool and doesn't dry out as much. You wet it and keep it wet. It helps a lot. The disadvantage is if the sun doesn't come out at any time (participant 5).

Next, participant 1 adds the benefit of reducing pests in agriculture caused by the presence of forest remnants on the rural property:

Places with more undergrowth are much less pest-ridden. I went to a plot of land in Dona Mariana and all around the lavoura is pure scrub. And the crops, you have to see how beautiful they are! It's sulphating very little. In our area here, it seems that the pests attack more. I think it's because of the lack of forest. If everyone was aware that those who have more land could let go a little, it would help. But the problem is that people aren't aware of this (participant 1).

At another point, the same participant defends the idea that forests are important for regulating the hydrological cycle, favoring infiltration and preventing surface runoff:

23Referring to pesticides. This speech is analyzed within the context: a way of farming that is less harmful to the environment, on page 106.

[...] The forest stops the rainwater. It rains there and stops there. So it keeps the soil wetter. If everyone cleans up too much, it will rain and the water will all go into the river. That's the big problem. It's not saying that it increases the water, it stops the rain there (participant 1).

Participant 1 then agreed with participant 2, who added: "it stores it, right", in a clear reference to recharging water tables.

At this point, it is interesting to note that participant 1 highlighted three different benefits provided by forests on rural properties, even standing out among the others for his level of involvement with the subject and the clarity of his ideas. However, as we will see later, when asked about the adequacy of the law for small landowners, or about the difficulty in conserving protected areas, this same producer takes a very critical stance on the conservation and/or recovery of APPs and RLs on smallholdings.

Although some producers didn't comment on the subject, in general it was clear that the importance of forests is known by the participants, and many of them identify that there are benefits even for agriculture.

5.2.3 <u>Adapting environmental legislation to small landowners</u>

The central ideas put forward by producers in the question: "Is environmental legislation suitable for small landowners?" are summarized in the following cognitive framework:

Figure 9. Cognitive framework: adequacy of environmental legislation for small landowners

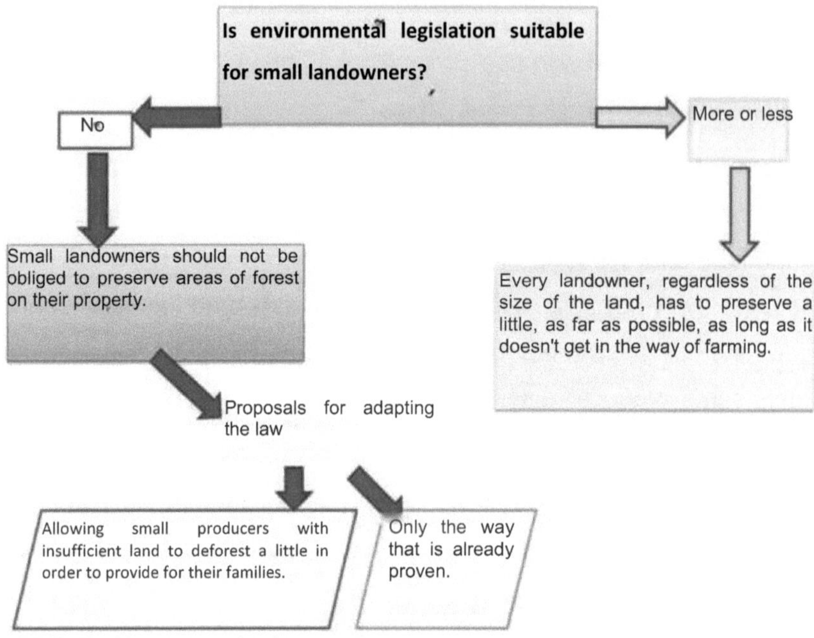

This question led to the group's first disagreement.

Participant 1 was the central articulator of the idea that smallholders should not be obliged to preserve:

Those who have small properties already have little to cultivate, so it gets complicated. There's a type of person who has a lot of land and nothing on it. I agree that there should be a law like this, that from 10 bushels upwards it should be compulsory to have at least 1 bushel of forest. The guy who has a bushel grows everything. If he lets go, how will he manage? There's no way (participant 1).

In this speech, participant 1 argues that the obligation to preserve should be attributed to landowners with more than 10 alqueires. It's important to note that a bushel in Rio de Janeiro represents 2.72ha[24] . Therefore, 10 alqueires is 27.2ha. Of the 83 properties analyzed in this study, only two were over this

---

24Some regions in the state of Rio de Janeiro use the São Paulo alqueire as a unit of measurement: 2.42ha or the Minas Gerais alqueire (alqueirão): 4.84ha, but the measurement commonly used in the mountainous region is the Rio de Janeiro alqueire of 2.72ha.

size. The average size of the properties analyzed was 7.5ha, approximately 3.6 times smaller than the proposed 10 alqueires.

He was then supported by participant 4, who repeated the same arguments and complemented them by highlighting the issue of family size and the division of land by heredity:

It's difficult for a rural family to have fewer than four children, especially back home. Nowadays it's less. A property with 5 bushels for a family of 5 to work is not much (participant 4).

The small landowner has no way of creating an environmental area, how can a guy with 1 bushel of land create an environmental area? Today, a guy with 1 bushel of land has 3 or 4 children to work with. Now, a guy who has 10 bushels upwards, then it's too easy to do an environmental area. For example: take away 1 and you get 9 (participant 4).

Sumidouro and the mountainous region have been reducing the size of plots of land, do you know why? Five or six years go by and the heirs increase. So, for example: you have 20 hectares of land, you have 10 heirs, it goes down to two. It's getting smaller, that's why the forests are running out (participant 4).

Participant 2 then gave a dissenting opinion:

I think that every piece of land, regardless of size, should have a bit of forest, because it's important not just for us, but for everyone in the watershed. That little bit that everyone can have is very important, as far as possible, as long as it doesn't get in the way of the crops [...] (participant 2).

Although participant 2 believes that everyone should preserve, regardless of the size of their property, she emphasizes that preservation should take place as far as possible, so that it doesn't get in the way of farming. In another statement she makes later on, it's clear that she made this reservation with the intention of forging self-defense:

Just like at home. Our land is very small. Then there's the pasture. The cow has to give the children milk. If you plant in front, where will the cow graze? There's a few meters of forest ahead, but the place to plant is near the river. If you pull it up 6 meters, you almost won't be able to plant (participant 2)[25] .

According to this report, it is clear that her site has a patch of native forest,

---

25This was reported at another point in the focus group, when asked what the producers know about the New Forest Code.

but that she is still in debt on the issue of APPs in riparian forests. The producer is aware of this, but apparently thinks it's unfair to have to dispose of an arable area on her site when she already has a preserved area.

Participant 1 then takes the floor again and defends himself by saying that he thinks it is important to have forest on the property, stressing that he does, but that he is speaking in defense of those who have very little land and whose livelihoods are under threat.

I think this law is right. Just like me. I'm going to preserve a small forest by the Rural River, but it's because I can. There are many people who won't have a place to work if I do this, but there's no way I can farm in a place like that (participant 1).

Saying that he thinks the law is correct at first seems contradictory, given the position he defended a few moments earlier. However, by analyzing the general context of his speech, we believe that what he meant was that he recognizes the importance of forests and that he thinks it's right to preserve areas that aren't useful for agriculture.

Next, participant 8 puts forward the idea that what should be charged is only the maintenance of the areas that are already conserved:

I don't think we can do away with what we have. You have to preserve what you have at least (participant 8).

Participant 7 tentatively agreed with this suggestion, but participant 1 disagreed and took the floor again, arguing that the law should even allow small landowners to clear land according to their need to increase the agricultural area:

I think the law should look at it like this: he has a piece of land and he sees that he can clear a bit of bush to increase his crops. Then the law had to allow him to do that, but come and look at it properly. Oh! You're going to take it away so you can plant it with your children. The law had to work like that. It should be valid for one type and another (participant 1).

Participant 4 agrees with participant 1. Next, participant 5 disagrees with the arguments put forward by participant 1, demonstrating her perception that the proposal that only landowners with more than 10 alqueires should be preserved would leave out most of the farmers in the watershed:

I think that if they give people who have little land a hard time: don't leave forest, don't preserve springs, don't do anything; out of ten people, one has a lot of land, nine have little. If these nine don't leave this huge place here, there will be about 10 preserved properties alone. And everyone who has a small property, which is the majority, won't do anything. So, in this case, I'm not saying as much as the big landowners, but the small ones have to do something. Because the majority are small (participant 5).

Participants 3 and 6 seemed to agree with participant 5, but did not add any new arguments. Table 5 below shows a summary of the participants' positions on this topic.

Table 5. Participants' opinions on the suitability of environmental legislation for small landowners

Idealized proposals:	Opinion held by:
Small landowners should not be obliged to preserve areas of forest, and can even clear them if they deem it necessary.	Participants 1 and 4
The small owner should only maintain what is already preserved.	Participants 8 and 7
The small landowner should preserve a little, as long as it doesn't get in the way of farming.	Participants 2
The small owner must preserve, but not as much as the large one.	5, 3 e 6
Did not speak	9

*It was not possible to identify whether this opinion fully agrees with the locations and metrics defined by the New Forest Code for preservation.

It's curious to note that in this question the participants put forward their proposals for how the law should be in order to defend their own wishes, but once again there was no mention of the changes brought about by the new law. Participant 2, by stating that she is dissatisfied with the size of the riverbank she will have to preserve, showed that she is familiar, albeit minimally, with the rules of the new Forest Code, but nothing was said about the old Forest Code, which demanded much more from small landowners.

5.2.4 <u>Problems caused by preserved APPs and RLs according to interviewees</u>

According to what the participants said, it was possible to identify two central ideas regarding the problems they identified with maintaining APPs and RLs in certain places on their properties, as can be seen in table 6 below:

Table 6. Participants' opinions on the problems caused by keeping APPs and RLs in certain places on the property.

Problems identified by participants:	Opinion held by:
Preserving the riparian forest is dangerous in the event of a flood.	1,2,4,7,8 e 9
Some trees in the forest consume the fertilizer from the crops.	1,2,3,4,6

Once again, participant 1 was the first to speak. This time he defended the idea that preserved riparian forests represent a danger to the population, based on the events of the tragedy in the mountain region in January 2011:

Rio, he makes his own way. Just like in this tragedy. What made the whole mess was the wood that came in with the water. Only those who were there saw it go by. Every tree trunk passed by that a car couldn't carry. I mean, it was uprooted from the riverbank. So there's no point in saying: I'm going to let a native forest grow on the riverbank. It will bring danger to Sumidouro, to Campinas. Because the day it floods, that wood will come. It's going to come tearing down. Then people think that a stick on the riverbank is an advantage. No! It's a danger to the population. Because, boy, in Campinas the bridge is blocked.

The water went into the street and over the roofs of the houses. If there were no sticks, the water wouldn't stop. So I mean, there's a project that defends the riverbank[26] , I agree, but creating sticks on the riverbank isn't good for the town, because when the flood comes, it uproots everything (participant 1).

Participant 4 shared the same view as participant 1:

Forest by the river will only hold more garbage, just like what happened a few years ago (participant 4).

Participant 7 implies that he thinks it would be better to delimit recharge areas

---

26Referring to Rio Rural.

for preservation rather than riparian forests. Then participant number 1 takes up the word again, giving a better example of the idea introduced by participant 7, and participants 8 and 9 agree with the suggestion:

It's better at the headwaters, right? (participant 7).

Up on the hill there has to be a forest there, but not on the riverbank. Just like I'm telling you. That was a total tragedy. It was a "passing" of sticks breaking everything, taking everything. A horror!" (participant 1).

The tragedy that took place in January 2011 had consequences for MBH Campinas that remain in the memory of the producers. Indeed, many homes and crops were destroyed by the flood, but none of the producers were able to look at the situation from the opposite perspective. After all, the houses and crops that were destroyed were where they shouldn't have been: on the banks of the rivers. This view presented by participants 1, 4 and 7 is in line with the figures obtained in the quantitative survey of this research. The data shows that 77% of the properties analyzed have a river or stream running through them and 81% of these properties have no riparian forest.

The preference for protection on hilltops is easy to understand, after all, the headlands are more difficult to access, which often makes growing olives more costly. This predominance of forest fragments in the recharge areas and the absence of riparian forests is easily seen when we analyze the satellite images, as we can see in figure 13 (a) and (b).

Figure 10. River bank and recharge areas in MBH Campinas

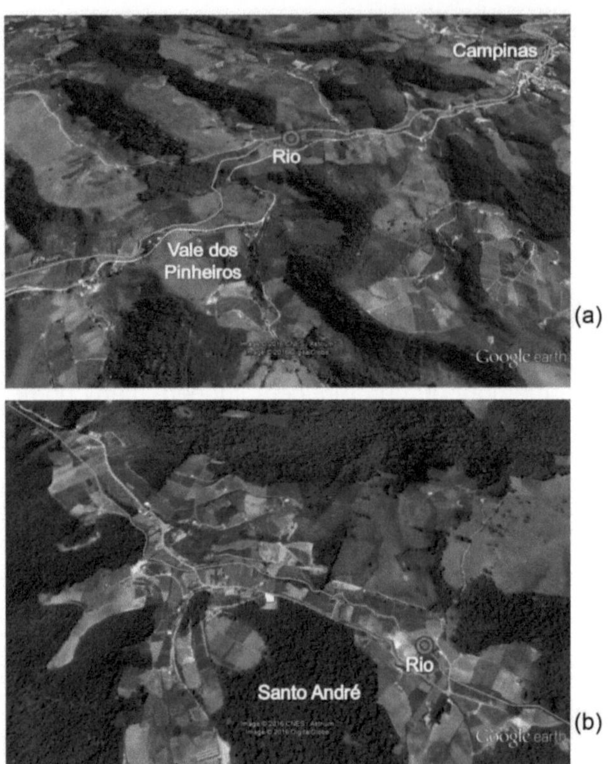

Caption: (a) and (b) satellite photos showing the main river of MBH Campinas, the small amount of riparian forest fragments and the existence of forest fragments on hilltops (recharge areas) that are more present. Source: Google Earth Pro.

As can be seen in these images, the banks of the rivers are heavily used for crops, while at the headwaters, where there are no pastures, it is common to find some forest fragments. The new Forest Code does away with the recharge areas through the rule of consolidated areas, however, according to article 61, paragraphs 1 and 2, the riparian forests of the MBH Campinas, even if they are consolidated areas, will still have to be recomposed with native forests on at least 5 meters of each bank on properties smaller than 1 fiscal module and 8 meters on properties between 1 and 2 fiscal modules. It is believed that this is the point at which the new legislation could cause the most changes in the MBH landscape.

According to Oliveira Filho (1994), riparian forests are extremely important for maintaining the quality of river water, reducing erosion on the banks, maintaining the ichthyofauna and improving the landscape. The new law makes it easier for producers, who will now have to give up a smaller cultivated area, to become regularized. However, many ecologists claim that 5 or 8 meters of forest may be insufficient to obtain the benefits described above. The participants in the focus group were not even sympathetic to these preservation strips reduced by the new Forest Code. It was possible to see that there is a myth that the ideal is to keep the banks "clean" for safety reasons and there are also those who don't want to give up an area that is being used for cultivation.

Some producers were also wary of maintaining preserved areas on their properties because they considered that some species could use the fertilizer from crops to grow, as we can see in the excerpt from the conversation below:

There's one type of tree that pulls more from the soil than another. There are trees that, from a distance of 20 meters from the crop, suck up the fertilizer. It takes the water, the fertilizer and doesn't let the crop produce (participant 4).

The eucalyptus doesn't allow it (participant 1).

Eucalyptus, you know. Eucalyptus really gets in the way. Just like on producer X's land. When we worked there, we even had to add more fertilizer (participant 2).

Participants 3, 6 and 9 agreed with the arguments put forward, but participant 5 intervened to point out that eucalyptus forests cannot be considered a conservation area:

But eucalyptus is not native forest, it's farming (participant 5).

It was possible to see that most of the participants in the focus group identify eucalyptus monocultures as a reserve area, even though they know that their exploitation has an economic purpose. Perhaps this is due to the sheer size of the plants, which possibly leads them to believe that this crop can have the same ecological effects as a native forest. From then on, the conversation took a different turn and it wasn't possible to see whether the criticism about the consumption of nutrients provided by the fertilizer also referred to native species.

5.2.5 <u>Causes attributed to the 2015 drought</u>

The central ideas put forward by the producers in the question: "What causes do you attribute to the drought of 2015?" are summarized in the cognitive framework below:

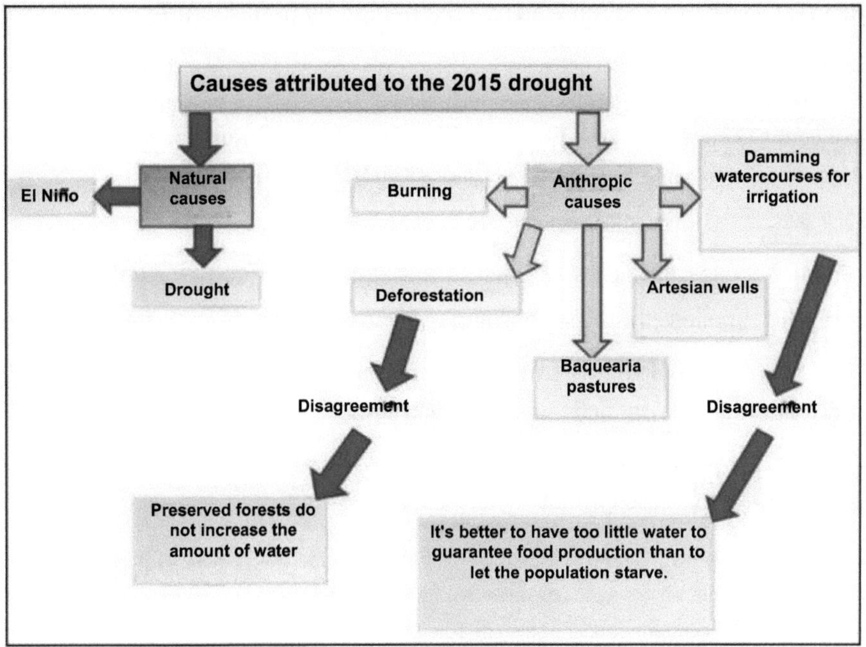

The first answers to the question "what causes do you attribute to the drought of 2015?" were related to natural phenomena:

Lack of rain, drought (participant 1).

At first, in our region, crops didn't even get wet. It rained a lot. Now, if it doesn't get wet, you don't even have to plant. Do you know what this is? They're saying it's an El Nino that's taking over (participant 4).

In participant 4's speech, he makes a temporal comparison, demonstrating that he realizes that the rains used to be more frequent. Later on, he attributes this climate change to the El Nino phenomenon. However, although the phenomenon did influence the rise in temperature and the reduction in rainfall in 2015, it is a cyclical event that can vary in intensity, but generally occurs at intervals of 2 to 7 years (OLIVEIRA, 2001). Therefore, this constant reduction in rainfall to which the producer referred may also be related to other causes.

The first to make the *link* between drought and human intervention was participant 2, whose speech was complemented by participant 7:

I think there are other things that get in the way of the rain coming down: fires. When the air is too heavy, it hinders the rain clouds. Too much heat, too much fire. People set fire to garbage in the yard. It goes into the woods. That gets in the way. In the old days we used to have to burn our garbage, but not nowadays. The truck comes to pick it up (participant 2).

Any little thing is set on fire (participant 7).

Participant 2 cites the issue of fires as a misguided strategy for getting rid of waste and implies that the burning of forests and fallow areas occurs accidentally on the properties. Up to this point, no direct relationship between drought and agricultural activity had been addressed, until participant 1 again decided to speak up, and his idea was then complemented by participant 4:

The other day I rode my motorcycle along the river and the water was getting in here (the farmer points to his shin), but it's not a lack of water. It's because it's dammed up. There's a lot of lettuce and tomato crops. There are people waiting in Soledade: One gets wet and it's over, and the other gets wet (participant 1).

In the state of Minas Gerais, if the water level drops below 30%, the engines are all sealed. No one can get wet there. Do you know why? Because it's the law in Minas Gerais. There you can't even spray Roundup[27] on the riverbank. It's the law there. The state of Rio doesn't have that. Does it? (participant 4).

In these speeches, the producers show great concern about the lack of water. Participant 1 highlights the district of Soledade in his speech because the region was most affected by the drought, due to the massive presence of lettuce and watercress crops, which consume an even greater amount of water than the lavouras commonly practiced in MBH Campinas. Participant 4, for his part, mentions a procedure for controlling the use of water for irrigation, which according to him is used in the state of Minas Gerais, but in his speech, he makes no mention of the source of this information. The most intriguing of the arguments he put forward was the perception that the use of

---

27Roundup is the trade name for a non-selective systemic herbicide manufactured by Monsanto, whose active ingredient is glyphosate.

glyphosate on riverbanks is only banned in the state of Minas Gerais. The lack of monitoring by the government and the lack of educational campaigns in the region may have contributed to this misconception.

Producer 8 then says in self-defense:

But if we don't plant and water, the people in the city won't have any vegetables in the market. If we don't water them, what will they live on in the city? It's better to have little water and food on everyone's table in the city [...] (participant 8).

Participant 8 recognizes that there is a problem of water scarcity due to irrigation in agriculture, but defends the idea that irrigation is necessary to guarantee food production. None of the producers addressed the waste of water in irrigation, which can be solved through simple measures, without the need to interrupt food production. No one mentioned the possibility of adopting drip irrigation, micro-sprinklers, installing a rain gauge or simply rationally sizing sprinklers.

Once again, comparing the focus group data with the quantitative questionnaire survey, we can see that the producers hardly depend on river water for domestic consumption. Only 2% of those interviewed claimed to use river water in their homes. However, rivers are very important for irrigation. In 40% of the properties analyzed, the water used to irrigate crops comes from streams and rivers. This is the second most common source among the producers interviewed. It is second only to springs by a small margin.

Taking into account that the participants generally consider it more appropriate to keep the riverbanks "clean" or with crops and that 45% of the properties analyzed do not have a septic tank, we can see that the participants are more concerned with the quantity than the quality of the water available in the rivers, perhaps because they don't use it for domestic consumption.

At another point in the focus group, when the question referred to the importance of forests, a discussion arose between participants 2 and 4 about the relationship between forests and rainfall:

[...] I think we should give priority to the Atlantic Forest, yes. Even the lack of water we've had there, we know it has a lot to do with the lack of forest (participant 2).

But does nature increase water? I don't think it does, you know why? The Caramandu region here in Sumidouro is the area with the most forest and yet the water has dried up. Do you know what's running out of water? Artesian wells. You make a well 300 meters deep. It's depleting all the water. There needs to be more supervision of this (participant 4).

Participant 4 explains his view that forests are not related to the volume of rainfall, arguing that even regions with a lot of forest were affected by the 2015 drought. In fact, deforestation is not the direct cause of the drought, but if there were more vegetation cover, the depletion of the reservoirs could be avoided.

He then presents the idea that deep semi-artisian wells are the cause of the water shortage. The big problem with semi-artisian wells is the risk of water contamination, since most of the time these wells are drilled without control by the competent authorities. However, the use of water located in these deeper regions has very little to do with the scarcity of available surface water (SILVA, 2010).

It was curious to see that later on, in another comment, participant 4 showed that he understood the principle of infiltration and surface runoff when he talked about brachiaria pastures:

I saw on Globo that a guy's spring has dried up near Vitória. All the guy's land is planted with brachiaria. Then a scientist came from outside, I don't know where, and said that he'd have to cut down the brachiaria for the water to come back. Because it rains and the brachiaria doesn't let the water go down to the ground. It goes straight into the stream. He'll sow fat grass to increase the water. Then I thought: in the past, all these hills were fat grass, now it's brachiaria. That's what's happening here, the brachiaria doesn't let the water enter the soil (participant 4).

This understanding could be extrapolated to native forests, just as participant 1 argued earlier[28] , but participant 2 doesn't seem to see the relationship between the two situations.

28First line on page 87.

5.2.6 <u>Ways of farming with less environmental damage according to the interviewees</u>

The central ideas expressed by the producers in the question: "Are there ways of producing food with less damage to the environment?" are summarized in the cognitive framework below:

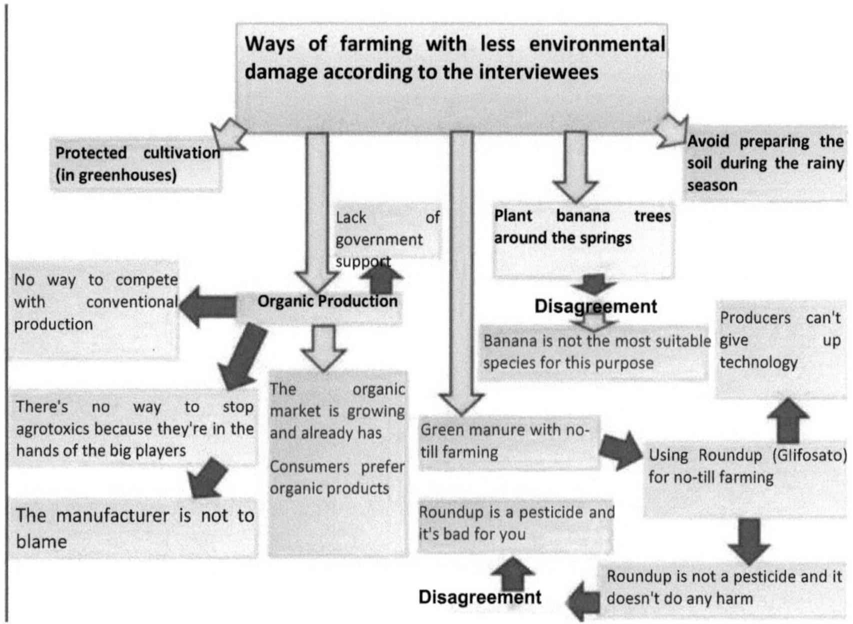

Figure 12. Cognitive framework: ways of farming with less environmental damage according to the interviewees.

The first proposal put forward on this issue concerned the control of soil preparation for planting in the rainy season:

You have to stop plowing in the summer. The rain comes and wipes everything out (participant 4).

These rivers are all full of dirt, but it's because there's no control over the "mining". They have to. After three or four months of rain, you can't touch the soil. It all goes into the river. This has to be controlled (participant 1).

According to the arguments presented in the speeches, it is clear that the participants are concerned about the silting up of rivers due to the risk of flooding. Nothing was said about the environmental impacts caused to the ecosystems directly involved in conventional soil preparation[29] .

Most of Sumidouro's streams are tributaries of the Paquequer river, which is used to supply the urban area. As well as the issue of land preparation

---

29Conventional soil preparation in MBH Campinas involves soil mechanization techniques with constant revolving of the surface layer every cycle, followed by the incorporation of lime and synthetic fertilizers.

carrying sediment and silting up the river, the pesticides used on crops contaminate fish and accumulate in the water chain, which can also cause serious problems for human health.

Still on the subject of soil preparation, in another speech, participant 4 brought up the proposal of green manure[30] allied to no-till[31] , but the producer also inserted another element into the discussion, which was the question of using a specific type of herbicide in the process:

We should do the same in Paty do Alferes. There, they drive a tractor over the land, sow something there, and when it's about that size, they apply Roundup and dig a hole to plant crops. Then there's no flooding. In the area of the environment, the product that does the least harm is Roundup, you know why? I apply Roundup to the soil, then a month later you lift the weeds and it's pure bait[32] from below [...]. Roundup isn't agrotoxic, because if it were it wouldn't bait the ground. But Gramoxone, Gramoxone[33] doesn't bait the ground (participant 4).

The proposal of green manure combined with no-till farming was very pertinent, however, the participant, in citing the municipality of Paty do Alferes as an example, seems to be unaware of the experiences with planting black oats promoted by the Rio Rural program in MBH Campinas. This technique has proved to be very efficient in maintaining the structure, humidity and organic matter of the soil, preventing the leaching of nutrients and the occurrence of erosion.

The producer's defense of the use of Roundup generates a separate discussion. Monsanto advertises that the product increases productivity while preserving the environment. The advertising even says that Roundup is the greatest partner of no-till farming. The product is sold in Brazil with a low

---

30Green manuring consists of planting leguminous or grass species that improve the fertility and structure of the soil so that the crop of interest can later be introduced to the same area.

31No-till farming is a soil management system that aims to reduce the impact of agricultural machinery on the soil. It is recommended to use the technique in conjunction with green manure. In this case, the green manure is only tilled in and the beds or pits are made without the straw being removed from the soil. In this way, a microclimate is promoted that favors the maintenance of organic matter and soil moisture.

32The producer refers to earthworms by using the term "bait".

33Gramoxone is the commercial name of a highly toxic non-systemic (contact) herbicide manufactured by Syngenta, the active ingredient of which is Paraquat.

toxicity classification (toxicological class IV), which gives producers the feeling that the product is safe.

According to Androli (2005), in the south of the country, where GM soya is produced, the positive image given to the product is even more intriguing. In his research, the author says that some farmers claim to have witnessed demonstrations proving that Roundup is not toxic to vertebrates: "sellers of the product would have poured the product into a bucket containing water and small fish and the result would have been positive, i.e. the fish remained alive".

The author adds that most studies on the effects of glyphosate and its derivatives on health and the environment are carried out by the product manufacturers themselves, who are interested in approving its use and boosting sales. In addition, it is very difficult for independent laboratories to carry out studies on Roundup, because the formulation of the herbicide and the products derived from it are protected by the principle of secrecy and industrial and commercial secrecy. Participant 4's view of Roundup is based on a comparison with another worse pesticide: Gramoxone, which is highly toxic to humans and other mammals.

While participant 4 was talking, participant 5 was restless and as soon as she had the chance she gave her opinion, disagreeing with participant 4:

I don't think there's any way that you can put something in the soil that doesn't allow weeds to grow and it be good for the soil. It's horrible. I think that because it doesn't allow weeds to grow, it must be one of the worst there is [...]. I don't think it's damaging the air, the water or anything like that, but it's damaging the soil too much. It should be banned (participant 5).

We can see that participant 5 didn't realize the indirect consequences of pesticide use when she said that the herbicide "isn't harming the air or the water". However, her suggestion that the herbicide should be banned was interesting and unexpected. Not least because she is also a conventional producer, and surely she or someone else in her family uses the product or an equivalent one.

Next, participant 8 argues in self-defense and brings a more reasonable view to the issue of prohibition:

Regarding this business of pesticides that don't allow weeds to grow: it's not the farmers who are to blame, but the companies that make them. Of course we're going to look for a way to benefit ourselves, but so that we don't harm nature, the manufacturer can't make it. Because of course, if they make it, someone will buy it (participant 8).

Then participant 5 takes the floor again:

The law should be strict and not allow manufacturing, because there was no such thing many years ago and that's not why anyone died of hunger or anything (participant 5).

At this point it is interesting to note that both defend the idea that, in order for the environmental impact not to occur, the ideal would be for the pesticide not to be manufactured. However, participant 8 starts from the perspective that it's not the farmer's fault, but the manufacturer's, and that as long as the pesticide exists he will continue to use it for his own benefit. Participant 5, on the other hand, showed that she attributed a certain amount of responsibility to the farmer and appeared to be concerned about environmental issues.

Next, the arguments presented by participants 1 and 4 corroborated those of participant 8 by showing a feeling of powerlessness in the face of the system they are part of:

There's no end to pesticides. Because pesticides are in the hands of the big players (participant 4).

In Rio de Janeiro, when I was 18, I went there to sell cargo. The cargo there was a pittance. Today you go there, my friend, there are no people, but there is a lot of cargo. Now, if nobody plants? If we go to the hoe, producing with the hoe will be bad. So we have to use technology. Thank God technology has arrived (participant 1).

Participant 1's speech shows concern about the issue of market competition for cargo volume. He shows a very restricted view of commercialization, seeing only the large metropolitan markets. He thanks God for the existence of technology, demonstrating his belief that without the package brought by the Green Revolution, he wouldn't be able to stay in the countryside.

Picking up on the idea that producers use agrochemicals to ensure their family's livelihood, participant 8 gives the following account:

Store X has the poison that kills the weeds. I'm there with 10,000 in debt at the store and I have my crop with the weeds smothering it, smothering it, squeezing it. There's often been a week of rain and I haven't been able to weed, and if I have 300 reais or 200 reais in my pocket I go and buy that medicine to apply to my crops, and a lot of the time it works. But if I offer someone 100 reais to weed, I need two or three days of service. He won't come. He already charges too much for two days [...]. So what do I do? I go there and buy a liter of poison that won't harm the crop and will save my crop, so that I don't lose it too. Save my debts, right? And also food for my family. And if the poison doesn't exist, then I won't buy it. Then it will be a problem for whoever made it (participant 8).

It is interesting to note the producer's criticism of the lack of manpower for services such as weeding. This phenomenon occurs in the municipality precisely because farming is predominantly family-based. People are busy with their own crops and there is a shortage of labor for these small jobs. In the logic of large markets, it is necessary to produce on a larger scale, after all the producer needs to have a volume of cargo to ensure competitiveness, but the producers apparently don't know how the marketing of organic products works, or products with added value, in which variety and quality are the great assets.

As for organic production, participant 2 said:

The organic issue. It's a great thing, but we have little access to it. There should be a law, the government should help us more here (participant 2).

There are a lot of people who buy vegetables with pesticides at the market because it's cheaper, but there are people who already prefer organic. I think agriculture needs to look at this (participant 2).

Participant 2 seems to be more receptive to dialog and learning about organic production. Like participant 5, who during the focus group showed a little more concern about the direction of agriculture in the watershed.

Nothing more was said about organic production or agroecology. After participant 2's speech, there were a few seconds of silence and then participant 1 introduced the proposal of protected cultivation:

There will come a time when everyone will have to produce food inside the greenhouse. Because the pests are attacking so much. It's poison and so on. If you produce inside the greenhouse, you can plant without the pests attacking (participant 1).

They are concerned about the occurrence of pests and the excessive use of pesticides (apparently because of the high cost). In some parts of the watershed there are some producers who are growing crops under cover in greenhouses. Generally, these producers are better capitalized and none of them work with the organic system. These producers end up being seen as examples of success by the others.

Another proposal suggested by participant 1 was to plant banana trees around the springs:

The banana in the water is number 1. It doesn't let the water run off. Whoever can plant it is very good (participant 1).

But participant 5 disagrees:

I've heard that this is a myth. Bananas aren't very good at increasing water. It's pure water inside (participant 5).

And participant 1 makes the rejoinder:

I say this from my own experience. Back home, there was a banana plantation up in a grotto. There was water there. That water was always there. It was dry, but the water was there. The banana plantation ran out and so did the water (participant 1).

Participant 1's proposal exposed an idea that is very common among producers in the watershed. There is a myth in the region that banana trees are suitable species for protecting springs, as we saw in participant 1's statement. In fact, the opposite is true. For banana cultivation, proximity to watercourses favours productivity, but a banana grove that is not intercropped with native species is not suitable for protecting springs, as it does not meet the principle of biodiversity and does not provide adequate structure to stabilize the soil.

The native vegetation around the springs also helps to maintain good water quality, acting as a filter for pollutants that may be present in the soil

(JUNIOR, 2009).

The quantitative analysis of this survey showed that springs are the main source of water for both irrigation (56%) and domestic use (56%). We also saw that 70% of the properties analyzed have one, two or three springs, but only 47% of those interviewed claimed that they have woodland around the outcrop.

An interesting proposal that would make it possible to combine the conservation of springs with the generation of income would be the consortium of banana trees with native species in the radius of the springs. Article 61, paragraph 13, item IV of the new forestry code states that the use of up to 50% exotic species of commercial interest is permitted, as long as the planting is interspersed. This would be a middle ground between what the producer proposes and the preservationist perspective.

Final considerations

The rules brought in by the new Forest Code will make it easier for properties in MBH Campinas to adapt, mainly because the majority of properties do not exceed 2 fiscal modules and fall under the rule of consolidated areas. In this way, the new FC will act much more to maintain what is already protected than to encourage the protection of new areas.

The restoration of riparian forests along watercourses is where the law will have the greatest impact on the landscape of the watershed if it is properly applied. However, even if producers recompose the area established by law, various studies show that 5 or 8 meters of protection is insufficient for the desired ecological effects. In this sense, this protection will act as a palliative, partially reducing the impacts of conventional agriculture on rivers.

It is believed that there will be a lot of resistance from landowners, even with the preservation strips reduced from 30 to 5 or 8 meters, as the participants in the focus group considered it inappropriate to maintain forest species on river banks due to the risk of flooding. In addition, the 20-year deadline stipulated by law for producers registered with the CAR to adapt is too long. During this period, it is not known whether the PRA will be able to help and motivate landowners in this endeavor.

The producers interviewed generally recognized the importance of forests, but most of them attributed the responsibility for conservation to large landowners. At many moments, the guarantee of food production and the survival of the rural population were used as a justification to legitimize the impacts of conventional agriculture - a discourse very much in line with that of the rural caucus and which shows how strongly the culture of deforestation is ingrained in the common sense of the producers. The low level of education of the public studied does not seem to be the main cause of this attitude, as there was no significant difference in the discourse of the few farmers with more schooling.

Bearing in mind that in family farming the conventional means of production and the process of occupying the countryside were established through incentives from the government itself, the idea that family farmers deserve differentiated treatment is fair. However, the way in which the changes were introduced in the new Forest Code is not a good thing, since technical and scientific concepts were ignored in the service of agribusiness interests. Similarly, this research showed that even with the loosening of the law, many of the interviewees were still resistant to their obligations. This attitude is based on the social reproduction of a model of cultivation that is incompatible with the conservation of forest remnants.

Therefore, it can be concluded that the implementation of public policies that make the agroecological transition viable on family farms is the main alternative to the problem at hand. These policies need to cover the entire production chain, guaranteeing credit, encouraging production and facilitating marketing. In addition, education and systematic monitoring of family farmers must be guaranteed, since the great challenge is to effectively raise awareness and this ideal only has full meaning outside the conventional system, in a logic where industrial capital does not dictate the rules of production systems.

References

ABRAMOVAY, Ricardo. *Paradigms of Agrarian Capitalism in Question.* Sâo Paulo: HUCITEC, 1992.

ALENCAR, G. V.; MENDONÇA, E. S.; OLIVEIRA, T. S.; JUCKSCK, P. R.; CECON, P. R. *Environmental Perception and Soil Use by Organic and Conventional Farmers in the Chapada da Ibiapaba.* Cearâ. RESR, Piracicaba-SP, Vol. 51, N° 2, p.217-236, Apr/Jun 2013 - Printed in July 2013.

ALENTEJANO, P. R. R. *A brief assessment of agriculture and agricultural policy in the state of Rio de Janeiro in recent decades. Rio de Janeiro, 2010.*

ASBRAER, Brazilian Association of State Organizations of

Technical Assistance and Rural Extension. *Rural Extension as an Essential Service to Brazilian Society.* Campinas, 2012. Available at: <http://www.asbraer.org.br/arquivos/noticias/ATERESSENCIAL-V5-Feliciano-Oliveira.pdf> Accessed on October 6, 2014.

ANDROLI, A. I. *Roundup, cancer and "green collar" crime.* Espaço Acadêmico Magazine, No. 51, August 2015. Available at: <http://www.espacoacademico.com.br/051/51andrioli.htm> Accessed on March 6, 2016.

ANTUNES, P. B. *Direito Ambiental.* 12. ed. Rio de Janeiro: Lumen Jures, 2009.

BARCELOS, E. A. S.; BERRIEL, M. C. *Institutional Practices and Interest Groups: the Geography of the Rural Caucus and Hegemonic Strategies in the Brazilian Parliament.* XIX Encontro Nacional de Geografia Agrária, Sâo Paulo, 2009.

BORDENAVE, J. D. *O que é comunicaçao rural.* 2 ed. Sâo Paulo: Brasiliense, 1985. 104p.

BARDIN, L. *Content analysis.* Lisbon: Ediçôes 70, 2000.

BARROS, C. BARCELOS, I. *The foresta traded on the stock exchange.* Available at:<http://apublica.org/2016/08/a-floresta-negociada-na-bolsa/> Accessed on November 1, 2016.

BOYD, H. W. J.; WETFALL, R. *Pesquisa mercadológica: texto e caso.* Rio de Janeiro: Getúlio Vargas Foundation, 1964.

BRAZIL. *Rural Environmental Registry.* Decree No. 7.830, of October 17, 2012.

BRAZIL. *Forest Code.* Law No. 12.651 of May 25, 2012.

BRAZIL. *Forest Code.* Law No. 4.771 of September 15, 1965.

BRAZIL. *Law of Guidelines and Bases of National Education.* Law No. 9.394 of December 20, 1996.

BRAZIL. *Provisional Measure.* MP No. 571 of May 25, 2012.

BRAZIL. *Environmental Regularization Programs for the States and the Federal District.* Decree No. 8.235, of May 5, 2014.

BRAZIL. *National System of Nature Conservation Units.* Law No. 9.985, of July 18, 2000.

BUAINAIN, A. M.; DI SABATTO, A.; GUANZIROLI, C. E. *Family farming: a regional focus study.* In: CONGRESS OF THE BRAZILIAN SOCIETY OF ECONOMY AND RURAL SOCIOLOGY, 42, 2004, Cuiabâ. Proceedings... Cuiabà: SOBER, 2004.

RURAL CHANNEL. *Government confirms extension of CAR for one year.*

Forest Code. Available at:

<http://www.canalrural.com.br/noticias/codigo-florestal/governoconfirma-prorrogacao-car-por-ano-56265> Accessed November 23, 2015.

CARNEVALLI, J. A.; MIGUEL, P. A. C. *Development of field research, sample and questionnaire for a survey study on the application of QFD in Brazil.* In: ENEGEP. XXI Encontro Nacional de Engenharia de Produçâo - VII

International Conference on Industrial Engine eringand Operations Management. Porto Alegre: SONOPRESS Indùstria Brasileira Represented by DISC PRESS Comércio Fonogràfico Ltda, 1999.

CARROL, C. R.; VANDERMEER, J. H.; ROSSET, P. M. *Agroecology*. New York: McGraw-Hill, 1990. 641p.

CARVALHO, H.; COSTA, F. *Peasant agriculture*. In: CALDART, R. *et al.* (Orgs.). Dicionàrio da educaçâo do campo. Rio de Janeiro: Escola Politècnica de Saùde Joaquim Venâncio, Sâo Paulo: Expressâo Popular, 2012. p. 26-32.

CECILIO, R. A.; REIS, E. F. *Didactic handout: watershed management.* Federal University of Espirito Santo, Center for Agricultural Sciences, Department of Rural Engineering, 2006. 10p.

Centro de Documentaçâo Histórica Pró-Memória - Secretaria Municipal de Educaçâo e Cultura - Prefeitura Municipal de Sumidouro/RJ. *Documentary collection.* Sumidouro, 2015.

CERVEIRA, R. & CASTRO, M. C. *Consumers of organic products in the city of São Paulo: Characteristics of a consumption pattern.* Informaçôes Econômicas, v.12, p. 7-20, 1999.

CONFEDERAÇÃO DA AGRICULTURA E PECUARIA DO BRASIL. *CNA claims 'complexity' and asks for CAR to be postponed.* Revista nacional da carne. 2015. Available at: <http://nacionaldacarne.com.br/cna-alega-complexidade-e-pedeadiamento-do-car/> Accessed on October 22, 2015.

COSTA, F. A.; CARVALHO, H. M. Campesinato. In: STEDILE, Joâo Pedro (org). *The agrarian question in Brazil: Interpretations of the peasant and the peasantry.* Sâo Paulo: Expressâo Popular, 2016.

CRACOLICI, M. F.; CUFFARO, M.; NIJKAMP, P. *The measurement of economic, social and environmental performance of countries: A novel approach. Social Indicators Research*, v.95, p.339-356, 2010.

DEAN, W. *A ferro e fogo: a história e a devastaçâo da Mata Atlântica brasileira.* 1. ed. Sâo Paulo: Cia. das Letras, 2004. 484 p. [1ª print 1996]

DELALIBERA, H. C.; WEIRICH Neto, P. H.; LOPES, A. R. C.; ROCHA, C. H. *Legal reserve allocation in rural properties: from cartesian to holistic.* Revista Brasileira de Engenharia Agricola e Ambiental, v.12, p.286-293, 2008.

DIAS, C. A. *Focus group: data collection technique in qualitative research.* Informaçâo & Sociedade: estudos, Joâo Pessoa, v. 10, n. 2, p. 2000. Ponto de Vista section. Available at:

<http://www.informacaoesociedade.ufpb.br/ojs2/index.php/ies/issue/view /35> Accessed on: 15 Feb. 2007.

DIEGUES, A. C. *O mito moderno da natureza intocada.* Sâo Paulo: Hucitec, 2001.

EGGER, D. S. *Continuities and ruptures: socio-spatial transformations in agriculture in Sumidouro*, RJ. 2010. 85f. Dissertation (Master's in Science) - Institute of Human and Social Sciences, Federal Rural University of Rio de Janeiro, Rio de Janeiro, 2010.

EMBRAPA. Brazilian Agricultural Research Corporation. *Agroecology: principles and techniques for sustainable organic agriculture.* Brasilia, DF: Embrapa Informaçâo Tecnològica, 2005.

EMATER-RIO. *Participatory Rural Diagnosis (PRD).* Sumidouro Local Office, Rio de Janeiro, 2013.

EMATER-RIO. *Annual Activity Report 2013.* Sumidouro local office, 2014.

EMATER-RIO. *Annual Activity Report 2014.* Sumidouro local office, 2015.

FAUSTINO, J. *Planificación y gestión de manejo de cuencas.* Turrialba: CATIE, 1996. 90p.

FEISTAUNER, D.; LOVATO, E.; SIMINSKI, A.; RESENDE, A. *Impacts of the new forest code on the environmental regularization of family farms*. Ciência Florestal, Santa Maria, v. 24, n. 3, p. 749-757, jul.-set., 2014

FERNANDEZ, F. *The attack on environmental legislation and the topicality of the tragedy of the commons*. O Eco, Rio de Janeiro, March 12, 2012.

FERREIRA, A. B. H. *Novo dicionârio da lingua portuguesa*. 2ª edition. Rio de Janeiro. Nova Fronteira. 1986. p. 1 787

FIDERJ, Institute for Economic and Social Development of Rio de Janeiro. *Studies for municipal planning: Sumidouro*. Rio de Janeiro, 1977.

FILGUEIRA, F. A. R. *Manual de Olericultura*. Vol. I. 2ª ed. Sâo Paulo: Agronômica Ceres. 1981. 338p.

FONSECA, M. F. de A. C. *Agricultura orgânica: regulamentos técnicos para acesso aos mercados dos produtos orgânicos no Brasil*. Niterói: PESAGRO-RIO, 2009. 119 p.

FONSECA, B. C. R. V. *As Principais Alterações Trazidas pelo Novo Código Florestal Brasileiro*. School of the Judiciary of the State of Rio de Janeiro. Rio de Janeiro, 2012.

FOSTER, G. *"What is a Folk Culture?"*, in American Anthropologist, vol. 55, n°2,1971.

FREITAS, H. *et al. The survey research method*. Revista de Administraçâo, Sâo Paulo, v. 35, n. 3, p.105-112, jul. 2000. Quarterly.

SOS MATA ATLÀNTICA FOUNDATION; NATIONAL INSTITUTE FOR SPACE RESEARCH. *Atlas of Atlantic Forest Remnants: 2005-2008*. Technical Report Sâo Paulo, 2009.

SOS MATA ATLÂNTICA FOUNDATION. NATIONAL INSTITUTE FOR SPATIAL RESEARCH. *Atlas of forest remnants, period 2012 - 2013*. Technical Report. São Paulo, 2014.

OSWALDO CRUZ FOUNDATION. INSTITUTE OF SCIENTIFIC AND TECHNOLOGICAL COMMUNICATION AND INFORMATION IN HEALTH.

Intensive consumption of pesticides in Rio de Janeiro reveals a scenario of "invisible" poisonings. Available at: <http://www.icict.fiocruz.br/content/consumo-intensivo-de-agrot%C3%B3xicos-no-riode-janeiro-revela-cen%C3%A1rio-de-intoxica%C3%A7%C3%B5es-> Accessed on April 05, 2016.

GARCIA, Y. M. *O código Florestal Brasileiro e suas alterações no Congresso Nacional.* Departamento de Geografia da FCT/UNESP, Presidente Prudente, n. 12,v.1, January to June 2012, p. 54-74.

GIEHL, G. *The general principles of environmental law. Portal Àmbito Juridico. 2007.* Available at:

<http://www.ambitojuridico.com.br/site/index.php?n_link=revista_artigos_reading&article_id=5083> Accessed January 09, 2015.

GLIESSMAN, S. R. *Agroecology: ecological processes in sustainable agriculture.* Trad. Maria José Guazzelli. Porto Alegre, UFRGS, 2000. 653p.

GROSSI, M. E. D. & SILVA, J. G. *New rural: an illustrated approach.* Londrina: Instituto Agronòmico do Paranâ. Vol. 1, 2002, 53 p.

HEREDINA, B.; PALMEIRA, M.; LEITE, S. P. *Sociedade e Economia do "Agronegócio" no Brasil.* Revista Brasileira de Ciências Sociais - Vol. 25 N° 74, 2010, p.159-196.

IBGE. Brazilian Institute of Geography and Statistics. *Agricultural Census 2006.*

IBGE. Brazilian Institute of Geography and Statistics, *Synopsis of the Demographic Census.* 2010.

IBGE. Instituto Brasileiro de Geografia e Estatistica, *Synopsis of the Demographic Census, Synopsis of Social Indicators - An analysis of the living conditions of the Brazilian population,* 2012.

INEA, State Environmental Institute. *Concepts for determining hilltop APP.*

Rio de Janeiro, 2011.

INESC, Institute for Socio-Economic Studies. *What kind of architecture is this?* Forest Code. Published on June 14, 2012. Available at: <http://www.inesc.org.br/biblioteca/textos/codigo-florestal> Accessed on November 29, 2015.

JUNIOR, C. S. *Avaliação de Projeto de Recuperaçao e Conservaçao de Nascentes no Municipio de Muzambinho-Mg.* 2009. 30f. Final Paper for the Higher Course in Coffee Technology - Federal Institute of Education, Science and Technology of Southern Minas Gerais - Muzambinho Campus. Muzambinho, 2009.

KLUCK, C.; REFOSCO, J. C.; CAGLIONI, E. and ARMENIO, G. A. *Impacto na economia das propriedades bananicultores em Luis Alves-sC, em função da implementação das áreas de preservação permanente. Rev. Arvore* [online]. 2011, vol.35, n.3, suppl. 1, pp. 707-716.

LAMARCHE, E. *Family farming: international comparison.* Campinas: Unicamp, 1997.2.ed.

LEFF, E. *Ecologia, Capital e Cultura - a territorialização da racionalidade ambiental.* Petrópolis: Vozes, 2009.

LEONARDO, H. C. L. *Soil and water quality indicators for evaluating the sustainable use of the Passo CUE river basin, western Paranà state.* 2003. 121p. Dissertation (Master's Degree in Forest Resources) - Escola Superior de Agricultura "Luis de Queiroz", Universidade de Sâo Paulo, Piracicaba, 2003.

MARTINS, S. V.; DIAS, H. C. T. *Importância das florestas para a quantidade e qualidade da água.* Açâo Ambiental, Viçosa, v. 4, n. 20, p. 14-16, Oct./Nov. 2001.

MARCHIORO, E. *et al. Application of the Brazilian Forest Code as a subsidy for environmental planning: a case study in the northwestern region of the state of Rio de Janeiro*. Soc. nat. (Online), Uberlândia, v. 22, n. 1, p. 11-21, Apr. 2010.

MARCONDES, D. *Forest, what forest for?* Available at <http://www.cartacapital.com.br/carta-na-escola/floresta-para-   quefloresta> Accessed on January 5, 2015.

MARCONI, M. D. A.; LAKATOS, E. M. *Técnicas de pesquisa: planejamento e execuçâo de pesquisas, amostragens e técnicas de pesquisas, elabor elaboraçâoçâo, anâlise e interpretaçâo de dados.* 3. ed. Sâo Paulo: Atlas, 1996.

MEDEIROS, R. *The Protection of Nature: from International and National Strategies to Local Demands.* Rio de Janeiro: UFRJ/PPG. 2003, 391p. Thesis (Doctorate in Geography).

MEDEIROS, R. *Singularities of the system of protected areas for the conservation and use of Brazilian biodiversity.* In: GARAY, I. & BECKER, B. (eds.) Dimensoes Humanas da Biodiversidade. Petrópolis: Editora Vozes, 2005.

MENDONÇA, S. R. *The National Plan for Agrarian Reform and rural employers' organizations in southeastern Brazil in the 1980s.* In: Second Conference on Comparative Regional History, Porto Alegre. Proceedings of the Second Conference on Comparative Regional History. Porto Alegre: PUCRS, PP. 1-20, 2005.

MINISTRY OF AGRICULTURAL DEVELOPMENT. DAP extract.

Search by municipality. Available at:

<http://smap14.mda.gov.br/extratopf/PesquisaMunicipio.aspx>   Accessed October 19, 2015.

MINISTRY OF THE ENVIRONMENT. *Ecological Corridors Project.*

Available   at: <http://www.mma.gov.br/areas-protegidas/programas-

eprojetos/projeto-corredores-ecologicos> Accessed on October 24, 2015.

MIYASAKA, S., NAKAMURA, Y. e OKAMOTO, H. *Agricultura natural*. 2. ed. Cuiabà, SEBRAE/MT, 1997.

MORGAN, D. L. *Focus group as qualitative research*. London: Sage, 1997.

MOSCA, A. A. O. *Hydrological characterization of two watersheds with a view to identifying hydrological indicators for environmental monitoring of planted forest management. 2003*. 96p. Dissertaçâo (Master's Degree in Forest Resources) - Escola Superior de Agricultura "Luis de Queiroz", Universidade de Sâo Paulo, Piracicaba, 2003.

MOURA, L. H. G. *Código Florestal Brasileiro: Elementos sobre a expressao ambiental da luta de classes no Brasil*. CAMPO- TERRITÒRIO: journal of agrarian geography. Special issue of XXI ENGA-2012, p. 1-25, June 2014.

MUSUMECI, L. *Small-scale production and the modernization of agriculture: the case of vegetables in the state of Rio de Janeiro*. Rio de Janeiro: Ipea/Inpes, 1987.

NEUMANN, P. S. and DIESEL, V. *The problem of not defining the basic economic unit in agriculture*. No 149560, 44th Congress, July 23-27, 2006.

NETO, O. C.; MOREIRA, M. R.; SUCENA, L. F. M. S. *Focus Groups and Qualitative Social Research: guided debate as a research technique*. FIOCRUZ/ENSP, 2002.

NEVES, D. P. *Family farming: how many moorings!* In: FERNANDES, B. M.; MARQUES, M. I. M.; SUZUKI, J. C. Geografia agrària: teoria e poder. Sâo Paulo: Expressâo Popular, 2007. p. 211270.

OKUYAMA, K. K.; ROCHA, C, H.; NETO, W.; ALMEIDA, D.; RIBEIRO, R. S. *Adequacy of rural properties to the Brazilian Forest Code: a case study in the state of Paranâ*. Brazilian Journal of Agricultural and Environmental Engineering. Campina Grande, PB, UAEA/UFCG, 2011.

OLIVEIRA. G. S. O *El Nino e Você - o fenômeno climàtico - Gilvan Sampaio de Oliveira*. Editora Transtec - Sâo José dos Campos (SP), 2001. Available at: <http://enos.cptec.inpe.br/saiba/Oque_el-

nino.shtml> Accessed October 26, 2016.

OLIVEIRA, L. B. *Family Farmers of Caramandù in Sumidouro (RJ): an approach to environmental perception and agroecology.* Darcy Ribeiro Pole - Volta Redonda, 2014. 62 f.

PEREIRA, A. M. C. *The logic of action in the reform of the forest code.* USP, Sâo Paulo, 2013.

PEREZ, F. & MOREIRA, J. C. *Health and the environment in relation to pesticide consumption in an agricultural center in the state of Rio de Janeiro, Brazil.* Cad. Saùde Pùblica, Rio de Janeiro, 2007, p. 612- 621.

PLOEG, J. D. V. *Seven theses on peasant agriculture.* In PETERSEN, Paulo. (org.) *Agricultura familiar camponesa na construçao do futuro.* Rio de Janeiro: ASPTA, 2009.

PORTO-GONÇALVES, C. W. *The globalization of nature and the nature of globalization.* Rio de Janeiro: Civilizaçâo Brasileira, 2006.

PORTO-GONÇALVES, C. W. *Os (des) caminhos do meio ambiente.* Sâo Paulo: Contexto, 1996.

PORTUGAL, A. D. *O Desafio da Agricultura familiar.* Agroanalysis magazine, March 2004.

REIS, L. C. *Revision of the Brazilian Forest Code: Impacts on the municipality of Bandeirantes - PR.* State University of Londrina. Londrina, 2011.

RODRIGUES, A. R. *Pontuaçôes Sobre a Investigaçao Mediante Grupos Focais.* Seminar COPEADI - Permanent Commission for Evaluation and Institutional Development, 1988.

RODRIGUES, V. A.; CARDOSO, L. G.; POLLO, R. A.; RE, D. S.; PISSARRA, T. C. T.; VALLE JUNIOR, R. F. *Morphometric analysis of the Ribeirao das Araras watershed - SP*. Revista Cientifica Eletrônica de Engenharia Florestal, Garça, v. 21, n. 1, p. 25-37, 2013.

SANTANA, D. P. *Integrated Watershed Management*. Sete Lagoas: Embrapa Milho e Sorgo, 2003. 63p. (Embrapa Milho e Sorgo. Documentos, 30).

SANTOS, Z. *A revisao do código florestal: Como se deu o debate politico durante a discussão sobre a alteração da Lei 4771, de 1965 - Código Florestal Brasileiro, na Câmara dos Deputados, como Casa Iniciadora - 1° ciclo de discussâo e votaçao*. Center for Formation, Training and Improvement of the Chamber of Deputies/Cefor. Brasilia, 2012.

SELBACH, J. R. *Social actors in conflict: The new Brazilian Forest Code*. UFRGS Porto Alegre, 2013.

FEDERAL SENATE. *Criticism of the concept of consolidated rural area*. 2012. Available at:

<http://www.senado.gov.br/noticias/Jornal/emdiscussao/codigoflorestal/aprovados-regras-claras-polemicas-area-rural-consolidada/criticas-aoconceito-de-area-rural-consolidada.aspx> accessed on October 22, 2015.

SILVA, C. A. *Manejo integrado em microbacias hidrogràficas*. Estudos Sociedade e Agricultura, November 3, 1994: 182-188.

SILVA, K. M. *The danger of indiscriminate use of artesian wells - A hydric approach*. Matias Barbosa - MG, 2010.

SILVA, R. R. *Sesmarias*. In: BiblioAtlas - Atlas Reference Library
    Digital Atlas of   Available   at
<http://lhs.unb.br/atlas/Sesmarias> Accessed August 20, 2018.

STOTZ, E. N. *Os limites da agricultura convencional e as razôes de sua persistência: estudo do caso de Sumidouro - RJ*. Rev. bras. Saùde ocup.,

São Paulo, 2012, p. 114-126.

SOUSA, I. S. F. *et al. Family farming in the dynamics of agricultural research.* 20p. Brazilian Agricultural Research Corporation, Ministry of Agriculture, Livestock and Supply. Embrapa Informaçâo Tecnològica Brasilia, DF 2006.

SOUZA, J. L. *Organic soil management: Emcaper's experience.* Viçosa, SBCS, v. 4, p. 13-16, 2000 (Newsletter).

TEIXEIRA, V. L. *Pluriatividade e Agricultura Familiar na Regiao Serrana do Estado do Rio de Janeiro,* RJ. 185 f. Dissertation (Master's Degree in Development and Agriculture) - Institute of Human and Social Sciences, Federal Rural University of Rio de Janeiro, 1998.

Teodoro, V. L. I.; Teixeira, D.; Costa, D. J. L.; Fuller, B. B. *The watershed concept and the importance of morphometric characterization for understanding local environmental dynamics.* Revista Uniara, v.20, p.137-157, 2007.

TOURINHO, L. A. M.O *Código Florestal na pequena propriedade rural: Um estudo de caso em três propriedades na Microbacia do rio Miringüava.* Federal University of Paranà - Curitiba, 2005.

VALERA, C. A. *Federal Law No. 12.651/12 - New (anti) Forest Code - An attack on sustainability and family farming.* Revista de geografia agrària. Special issue of XXI ENGA-2012, p. 1-17, jun., 2014.

VIGNA, E. *Bancada ruralista: The largest interest group in the National Congress.* Year VII n°12. Brasilia, INESC, 2007.

Appendix A

QUESTIONNAIRE TEMPLATE USED

<u>Socio-economic data</u>

1 - Name:

2 - Site size:

3 - Education:

( ) Higher Education   ( )High   School Completed

( ) Secondary school incomplete( )   Elementary school complete

( ) Elementary school incomplete ( )   Never attended school

4 - Use of water resources and sanitation:

Do you have a septic tank on your property? ( ) Yes (   ) No

5 - How do you collect the water you use to irrigate your crops?

( ) Spring   ( ) River or stream   ( ) Artesian well

6 - How do you collect the water you use at home?

( ) Spring   ( ) River or stream   ( ) Artesian well

<u>Protected areas on rural properties</u>

7 - Do you have any woodland on your property (except capoeiras and fallow areas)?

8 - The forest occupies an area that would fit how many tomato plants?

9 - Do you have springs on your property? How many? Do you have forest around it? (consider only native forests)

10 - Does a stream run through your property? Do you have forests on at least 80% of the banks, considering a minimum of 5 meters from the riverbed?

11 - Do you consider the area available on your site to be sufficient for your

activity?

Appendix B

SCRIPT FOR THE FOCUS GROUP

Key question

1 - Why have forests on a rural property?

• What benefits do forests bring to the environment?

• Can the lack of forests in a rural watershed negatively affect agricultural activity? Why?

• What causes do you attribute to this year's drought?

2 - Is environmental legislation suitable for small landowners?

- Do you know, even minimally, the part of the environmental legislation that deals with rural properties (contained in the New Forest Code Law 12.727/2012)?

3 - How difficult is it to conserve green areas on a rural property?

- Do forests ever get in the way of agricultural activity?

4 - Are there ways of producing food without damaging the environment?

More
Books!

info@omniscriptum.com
www.omniscriptum.com
OMNIScriptum

Printed by Books on Demand GmbH, Norderstedt / Germany